T0100118

Advance Praise for *Google Leaks*

"Rarely do we get to glimpse inside the big tech companies upon which we rely for information. Zach Vorhies is an American hero who sacrificed his livelihood to warn Americans that instead of providing its users with neutral information based on search engine traffic volume, Google employs the most powerful thought manipulation algorithms ever devised to steer its users to outcomes that favor the company's self-serving mercantile and ideological ambitions."

—Robert F. Kennedy Jr., bestselling author of *Crimes Against Nature*, *Thimerosal*, and *Framed;* senior attorney for the Natural Resources Defense Council; and president of Waterkeeper Alliance

"For years we analyzed data because we saw wild unexplainable shifts in many metrics. It was only after Zach came forward with the hard internal document that named our very channel were our suspicions confirmed beyond a shadow of a doubt. Suppression is real. Zach risked everything so that freedom of speech could survive. And for that America should be eternally grateful, because I am."

—Gary Franchi, founder of Next News Network

"Americans have always been taught that the threat to their liberties came from government. Zach Vorhies sacrificed his career to bring us irrefutable evidence that in twenty-first-century America, the threat to liberty—an unprecedented threat to liberty—comes from private enterprise, namely, big tech, which controls more information than any government except for those in totalitarian countries. Big tech increasingly censors and manipulates information like totalitarian governments do. Zach Vorhies' book is the needed wake-up call to prevent totalitarianism in America. Its importance cannot be overstated."

—Dennis Prager, radio host of *The Dennis Prager Show* and writer

"*Google Leaks* is a dramatic expose of Google's totalitarian mission to censor its audiences, disseminate propaganda, and decide election outcomes."

—Michael Rectenwald, PhD, author of *Google Archipelago*

"While most people would have turned a blind eye and ignored the blatant suppression of the First Amendment by the Google technocrats, Zach

Vorhies drew his line in the sand and stood on the side of truth, righteousness, and freedom of speech. *Google Leaks* uncovers the underbelly of the Orwellian censorship policies that are an integral part of Big Tech's propaganda machine (that would have made Edward Bernays proud). We are so thankful for Zach's conviction and bravery. He's an American hero, and this book will blow you away. Just read it. You won't regret it."

—Charlene Bollinger, filmmaker, *The Truth About Cancer*

"Zach's shocking revelations come at a crucial time for the world. At a time when the Oligarchy looks to control the world's information, we now know with certainty Google's plan for us, and it definitely isn't full of rainbows and butterflies."

—Ryan Hartwig, Project Veritas Facebook whistleblower and author of *Behind the Mask of Facebook*

"Zach Vorheis has dedicated his life to transparency, accountability, truth, and the facts, which has enhanced every person's Freedom & Liberty."

—Dave Janda, MD, social media influencer and Obama Care whistleblower

"Zach is a rare breed of human. His story of self-sacrifice and internal struggle is a story of an American hero. The choice he wrestled with—sacrificing his dream job to illuminate the evils of the world's most powerful corporation—will go down as a pivotal moment in history. The public started to shift its adulation of the once "Don't Be Evil" company to seeing it as behaving in an evil manner. We cannot thank Zach enough for the service he's done and is continuing to do for the public good and right to know. And I'm honored to know him and call him a friend and brother-in-arms."

—Cary Poarch, Project Veritas Whistleblower, #ExposeCNN

"Tech censorship has been the harbinger of doom for not only the public discourse but the livelihoods of political dissidents for over half a decade. As things get exponentially worse, it's easy to forget the human element to the censorship machine. Not everyone on the inside is without conscience. Zach Vorhies' story paints a compelling picture that reminds us that we needn't look only to politicians to fight the battle against big tech—a more immediate force can be found right under the oligarchs' noses."

—Zach McElroy, Project Veritas Facebook whistleblower

"Zach Vorhies' courage and service to humanity is inspirational! This is a man who could have chosen to sit back and collect a big paycheque working for one of the most powerful companies on Earth. Instead, he opted to forego his personal security in order to expose Google's manipulations to the world! I thank him not only for providing evidence of Google's power-hungry programmers, biased algorithms and censorship regime but also for inspiring others to come forward."

—(Amazing) Polly St. George, social media influencer

"What Zach brought forward changed the public debate on Big Tech and we at Project Veritas are grateful that we could be the vehicle for his bravery, knowledge, and conviction."

—Matthew Tyrmand, Board of Directors, Project Veritas

GOOGLE LEAKS

A WHISTLEBLOWER'S EXPOSÉ OF BIG TECH CENSORSHIP

Zach Vorhies and
Kent Heckenlively, JD

Skyhorse Publishing books may be purchased in bulk at special discounts for sales promotion, corporate gifts, fund-raising, or educational purposes. Special editions can also be created to specifications. For details, contact the Special Sales Department, Skyhorse Publishing, 307 West 36th Street, 11th Floor, New York, NY 10018 or info@skyhorsepublishing.com.

Skyhorse® and Skyhorse Publishing® are registered trademarks of Skyhorse Publishing, Inc.®, a Delaware corporation.

Visit our website at www.skyhorsepublishing.com.

10 9 8 7 6 5 4 3 2

Library of Congress Cataloging-in-Publication Data is available on file.

Print ISBN: 978-1-5107-6736-2
Ebook ISBN: 978-1-5107-6737-9

Printed in the United States of America

"It's not so much staying alive, it's staying human, that's important. What counts is that we don't betray each other."

—George Orwell

"Two decades ago, Google became the darling of Silicon Valley as a scrappy startup with an innovative way to search the emerging internet. That Google is long gone. The Google of today is a monopoly gatekeeper for the internet, and one of the wealthiest companies on the planet, with a market value of $1 trillion and annual revenue exceeding $160 billion."

—Opening of Department of Justice Anti-Trust Lawsuit filed against Google on October 20, 2020[1]

Contents

Foreword

by James O'Keefe

There are those in this life who don't give into threats and are not chasing after rewards. Zach Vorhies chose to follow his conscience, even if it led him past the gates of hell. He took on the most powerful company in the world, Google, from within. In doing so, he would not only educate people on Orwellian concepts of "algorithmic unfairness," the greatest legacy from his whistleblowing would be to inspire countless others to follow suit.

—James O'Keefe
Project Veritas

Prologue

San Francisco, California, August 5, 2019

Zach Vorhies
@Perpetualmaniac

Deadman's switch is now activated. In the event of my untimely death ... full public disclosure.

11:42 AM · Aug 5, 2019 · Twitter for iPhone

View Tweet activity

191 Retweets **52** Quote Tweets **394** Likes

Google wants me dead.

That was the only scenario that made sense when my friend called and said, "Zach, the police are here and they're looking for you."

I'm a big fan of the classic quote from Sun Tsu's book, *The Art of War*: "All warfare is based on deception."

That's why, in my "exit interview" from Google a few weeks earlier, I'd given them a phony address, that of my friend. They suspected I was the "anonymous whistleblower" who'd appeared in a *Project Veritas* video with James O'Keefe a few months earlier. In that interview, I detailed how I'd collected more than nine hundred pages of documents from internal servers, which I'd been legally permitted to do at the time as a Google employee. And I made it clear that Google was lying about their claim to be a "neutral platform" or a modern day "town square" where everybody was free to speak their mind.

I'd worked for Google for eight and a half years, become a senior engineer, and had loved the original company slogan, "Don't be evil," and was even okay with the slightly less dramatic replacement, "Do the right thing." But the company I loved had changed. In my youth, I'd read several dystopian science fiction novels such as George Orwell's *1984*, Aldous Huxley's *Brave New World*, and Ray Bradbury's *Fahrenheit 451*, in which the main character is a fireman whose job is to burn houses containing outlawed books. I felt like I was living one of those stories.

I'd read a fair amount of the political philosophers and become intrigued with Jeremy Bentham's idea of the "Panopticon," a system of control in which the inmates of a prison could all be observed by a single guard. Due to the design, none of the prisoners in their cells knew whether they were being watched. As a result, the inmates lived in fear and monitored their behavior with a minimum of oversight. It seemed like a good metaphor for Google's recent actions.

But free people are not supposed to be treated like prisoners.

And that's what I believe my former employer was doing with its censoring, algorithms, and other tricks designed to bring about what they believed to be a better world. But that's the claim every tyrant in history has made to justify their atrocities. My soul was sickened by these underhanded tactics, and I didn't want to be a part of it anymore.

As my friend told me about the actions of the police, I could see the plan in motion. I can have a hair-trigger temper when I believe something is wrong. My mouth can sometimes run ahead of my brain. I believe Google was counting on that.

They expected me to do something stupid.

But with age comes wisdom, and instead of panicking I slowed down my brain. *Think it through, Zach,* I said to myself. *The police are looking for you, but you've committed no crime.*

Everything I'd learned in my life would determine whether I made it through that day without winding up in trouble with the law, in jail, or dead.

CHAPTER ONE

Google Turns toward the Dark Side

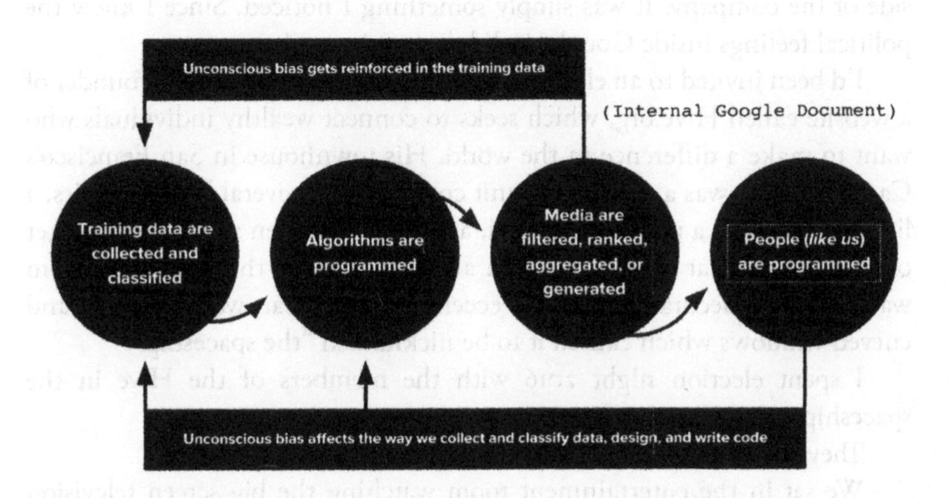
Unconscious bias gets reinforced in the training data
(Internal Google Document)
Training data are collected and classified
Algorithms are programmed
Media are filtered, ranked, aggregated, or generated
People (*like us*) are programmed
Unconscious bias affects the way we collect and classify data, design, and write code

On November 8, 2016, I was working at Google's YouTube office in San Bruno, California, deep into a programming project, when in the late afternoon I noticed a co-worker walk by with a scowl on his face.

I went back to my work, but after a few minutes became curious. I got up and went into the TV room to get some coffee. The room had several screens, tuned to different networks, and I saw there were many people in the room. There was a depressed vibe, as several people had a hand to their mouth as if in horror, while others were resting their tilted head on a palm as if in defeat.

I said, "Hey, guys! How's the election going?"

One guy spoke up. "Not good."

"What? Is Trump winning?"

Another guy responded. "Yeah. Big time. I think he's going to win it."

"No way," I said. "Clinton's got it in the bag."

It wasn't that I liked Clinton. In fact, I didn't. I was appalled by what she'd done to Bernie Sanders in the primaries that year. And I'd read enough about the "Clinton body count" to suspect she was probably guilty of some pretty shady stuff. I was aware the leftists hated Trump so much, which made me mildly amused if I thought about it. But honestly, I didn't spend much time thinking either way about it. It also seemed clear to me Google was boosting positive Hillary Clinton news to the top of the search results, while at the same time, boosting negative Donald Trump stories. Later it came out Facebook was doing the same thing, so it didn't really surprise me to see it at Google. Nothing I was working on put me in contact with that side of the company. It was simply something I noticed. Since I knew the political feelings inside Google, it didn't surprise me.

I'd been invited to an election night party at the house of the founder of a website called Hive.org, which seeks to connect wealthy individuals who want to make a difference in the world. His townhouse in San Francisco's Castro District was a multi-level unit connected by several levels of stairs, a living room with a pool and Jacuzzi, an enormous open space at the center of the unit, and at the top level an amazing view of the city. Each room was colorfully decorated, fun and eccentric, with clean wooden floors and curved windows which caused it to be nicknamed "the spaceship."

I spent election night 2016 with the members of the Hive in the spaceship.

They were not a happy group.

We sat in the entertainment room watching the big screen television broadcast the returns. Some people were stunned. Some were shocked. And others were crying.

Trump gave his victory speech, and then instead of Hillary coming out to make a concession speech, all we got was her chief of staff, John Podesta. I heard a couple people mutter, "Oh, my God, is Hillary drunk in her hotel room?"

I heard one person say, "If I lost to Trump, I'd get drunk, too."

The past few years had been very stressful on me, personally, professionally, and financially. I'd become so politically apathetic after watching the failure of the Occupy Wall Street movement, as well as the suspicious rise of Antifa, that I just didn't know what to think of it all. I understood everybody around me was super depressed, so I didn't make any comments either way. I figured I'd just go home to bed and try to figure it all out in the morning.

* * *

When I woke up the next morning, the first thing I said to myself was, "Hillary Clinton's not going to be president." I paused for a moment to let the enormity of the realization sink in. Like most everybody else around me, I was convinced she'd easily win. The next thing I said was "Donald Trump is going to be president."

Then I just started laughing. I didn't realize how much I'd been dreading the prospect of a Hillary Clinton presidency. From my readings about the "Clinton body count," I had some significant suspicions about the former secretary of state and her husband. And now, the guy who'd been saying he was going to "lock her up" and suggesting Barack Obama was born in Kenya was going to be president? It was all just so hilarious. A "conspiracy theorist" (a term invented by the CIA to disparage those looking into the Kennedy Assassination[1]) was going to be president of the United States.

I continued laughing to myself that morning, as I took my shower, ate my breakfast, and walked to the street where I caught a Google bus to take me to YouTube. I laughed all the way to work, thinking how it was going to trigger so many leftists at the company. You might think I was acting like a troll, but I was getting so tired of the leftists who'd won every single battle in the culture wars for the past thirty years, and it still wasn't enough.

I knew those people. I'd protested with them. I'd worked alongside them for years.

They'd just had a black president for eight years. Then because America had picked a white man instead of a white woman, we'd suddenly changed into a racist country? The leftists I'd come to know, through protesting, and at work, were never satisfied. The demands were always more, more, more, and their proposals became increasingly more radical. These bullies just got a bloody nose from somebody who played the game just as hard as they did.

Employees at Google were taking the day off because they were having emotional meltdowns over the election, as if a close family member had died. I can't tell you the number of people I saw at work crying, having pained conversations with their fellow employees, or offering hugs to people who just couldn't deal with the situation. Did these people have even a basic understanding of civics? There are these things called elections. You try to win these "elections" by appealing to more voters than your opponent.

Sometimes you win and sometimes you lose.

When you lose, you take a long, hard look at your campaign and say, "What did the other person do better than me?" Then you change strategies, tactics, and sometimes even your positions, so that next time you can win. Americans had been participating in these contests since the election of George Washington in 1789. Before 2016, I'm pretty sure no company ever gave a "mental health day" to deal with the results of an election.

Initially I was hopeful that the Google employees would start asking the questions which normally followed an election loss. Questions such as "What did we do wrong?" or "How can we do better next time?" But it quickly became apparent no such reckoning would take place. Instead the halls were quickly filled with talk of how the election "wasn't fair," that electors must be encouraged to change their votes, and that a "resistance" to Trump needed to form.

But all that was simply a prelude of what was to come.

An "All-Hands" meeting took place at Google's corporate headquarters in Mountain View, CA on Thursday, November 17, 2016, which would set the stage for the company's future actions.

* * *

Officially, these "All-Hands" meetings were called TGIFs (Thank God it's Friday). They had originally been on Fridays, but as our Japanese office became quite large it was decided to hold them on Thursday afternoons so the Japanese employees could watch as well.

Since I was working at the YouTube office in San Bruno, I watched the meeting on my desktop computer. Google was sending the feed out to all its employees. The Google main headquarters in Mountain View is known as the "Googleplex" with two million feet of office space. It also boasts a replica of Spaceship One, which won the Ansari Prize in 2004 for being the first privately funded crewed rocket, a life-size replica T-Rex skeleton, nearly thirty restaurants, a bowling alley, a sculpture garden with larger than life-sized emojis, such as eight foot high frosted donuts, and seven fitness centers, in addition to yoga classes and massages.[2]

I want you to imagine the main auditorium at Google, filled with hundreds of employees, many of them wearing the trademark Google hat, a multi-colored beanie with a propeller on top if they were new. This is referred to as "Googley-ness" and in normal times denotes an off-kilter sense of humor. But with the Trump election and their despair over the results, it seemed to morph into something much different.

Sergey Brin, the co-founder of Google and CEO of Alphabet, the parent company of Google, appeared on our screen. He was forty-three years old at the time, slender, standing about five-foot-eight, his wealth in 2016 estimated at thirty-five billion dollars, dressed in a blue, casual long-sleeved shirt, shaggy black hair, beard and mustache, reminding me of a scruffy 1970s Al Pacino from a *Serpico* poster. He was relaxed and comfortable with himself, like he might be the senior computer engineer in the next cubicle, or the team leader everybody loved.

"Okay, folks," he began. "I know this is probably not the most joyous TGIF we've had. And you know, let's face it, most people here are pretty upset and sad because of the election. But there's another group, a small group that we should think about, who are very excited about the legalization of pot."

Laughter, applause, and whoops came from the audience and Sergey's mood seemed to lighten. A small, mischievous smile played across his lips. "I was asking if we could serve joints [marijuana] out on the patio. Apparently, these things take a little while to take effect. A huge disappointment. I've been bemoaning that all week. I'll be honest with you."

He paused for a moment, letting the laughter pass:

> On a more serious note, myself, as an immigrant, *I certainly find this election deeply offensive*. And I know many of you do, too. And I think it's a very stressful time. And it conflicts with many of our values. I think it's a good time to reflect on that, and hopefully we're going to share some thoughts today.
>
> I guess there are two dominant reasons to be upset. One is because so many people apparently don't share many of the values we have. I mean, I guess we've known that for many months now. It's not like, in election terms, whether it was 47.2 percent or 48.2 percent, whatever it was. It's always been a lot of people that feel that way, apparently. But confronting that is pretty upsetting.
>
> And secondly, confronting the reality of an administration that's now forming. And look, we have no idea what it's going to do. That's the honest truth. We have no idea what direction the country will take, whether the past policy proposals are serious or whatnot, it's a period of great uncertainty.
>
> And for many of us here, especially immigrants, minorities, women, so many people. And it's just generally people who have kids and wonder about

> their world. So, I don't have great answers for you, up here today. But I think it's important we chat about it and are thoughtful about it in the coming months.
>
> And with that, Sundar . . .

* * *

Sundar was Sundar Pichai, who'd been named CEO of Google on August 10, 2015, when it was reorganized under the parent company, Alphabet. Sergey was president of Alphabet, the new parent company of Google, and Sundar served under him, although it was always portrayed as a collaboration. A few inches taller than Sergey, Sundar was wearing Silicon Valley casual: tennis shoes, jeans, and a light grey zip-up cotton hoodie. Unlike the often shaggy and disheveled look of Sergey Brin, Sundar was a trim man, well-manicured with close cropped black hair and beard, glasses, and his cultured Indian accent spoke of both intelligence and thoughtfulness.

Pichai took the stage, nodding to Sergey. "Thanks for that," he said to a retreating Sergey, then looked out at the audience. "It's good to see all of you here. I'm glad we're getting together at a moment like this. It's been an extraordinarily stressful time for many of you. The outcome in a two-party system, with a lot of polarization in the country, a deeply divided country, and you have a binary outcome, right? There is no easy way through this. And historically all political processes are stressful and tough, particularly if the outcome is not what you hoped for."

I was starting to get a little concerned as to where the conversation was heading. They were acting as if elections ALWAYS went the way they wanted, and if they didn't, the event should be treated like a death in the family. *No easy way through this?* Were Google employees braving a hail of bullets to land on an occupied beach? Were they running into a burning building to save children?

Sundar continued:

> "On top of that, I think we'd all agree this election was particularly hard. There was a lot of rhetoric. And there were a lot of groups targeted. And I think that makes it a very hard cycle, especially with our values.
>
> But, I hope, a couple things I'd say. It's important to remember we're in a democratic system. And it's heartening to see a transition happen properly. I grew up in India and there were a lot of things wrong. But it was a democratic country. And we've gone through many, many, many hairy moments like

> this, right? It was a poor country of one plus billion people going through a democratic process with many more divided opinions that what you're seeing here. And I've seen over time, have faith in it. It tends to work out. There are many, many scary moments and it looks like the wheels are coming off. But it tends to make it through okay. And it seems to be better than any other system out there.
>
> I think we should keep that in mind. I think it's a good moment of reflection, introspection, and listening to each other, too. I think part of the reason the outcome ended the way it did, is people don't feel heard across both sides. And I think it's important to reach out and talk to each other. There is a lot of fear within Google and I've gotten a lot of emails, to my notebook. And I would tell most Googlers there are people who are very afraid.
>
> Sergey pointed out the many groups, women, blacks, people who are afraid, based on religion. People who are afraid because they're not sure of their status. The LGBT community and I could go on. There's a lot of fear. And so, I think it's important to reach out, be aware of that fear. Be sensitive and try and talk and have conversations. We are so deeply committed to our values. Sergey mentioned that at the start. Nothing will change. We will always stand up for the values we believe in. And especially in a society you stand up for people who are minorities. And that defines us, and our country needs to do that.
>
> I think we have a few more people who are going to come and say their thoughts. I'm not sure I can get through everything I want to say. So, I'm going to have Kent [Walker] come, and say a few things, and we'll come back. We have a few more people who want to say their thoughts.

* * *

As Sundar finished his talk, Kent Walker, vice president for Global Affairs at Google and chief legal officer, took the stage. Like the two previous speakers, he wore tennis shoes and jeans, but instead of a hoodie or long-sleeved, crew neck shirt, he wore a blue, button down dress shirt, although no jacket, perhaps a nod to his legal education, but expressed in Silicon Valley fashion style. Walker, a graduate of Harvard University and Stanford Law School, was similarly trim as the other speakers, but looked more Middle-American with his thinning brown hair and prominent forehead. Put him in a typical thousand-dollar attorney suit and he'd look completely at home in any large city law firm.

Sundar handed over the microphone to Kent, and he began pacing the stage:

> So, look, it was a shock to all of us, the results of the election. It was a fair and democratic process and we honor that. But at the same time, it showed an incredible level of division among Americans. And that's something that gives us pause. And focuses us on, how did we miss that? What can we do to reach out to people whose perspective we have a hard time understanding?
>
> But it's not just a challenge for America. It's a challenge that goes well beyond America. The implications for the rest of the world are vast. And the echoes around the world are significant. This is not the first time that we've seen this rising tide of nationalism, populism, and concern. There are drivers of globalization and immigration which have sparked movements throughout Europe, throughout Asia, throughout Latin America. It's not just Brexit, but rising new parties are coming onto the scene, splintering traditional parties. Through Germany, France, Italy has a referendum next month, the Philippines, Thailand, big chunks of Latin America.
>
> We're trying to figure out our right next steps in that. But we recognize that globalization and the internet have been an incredible force for change. They have brought hundreds of millions of people out of extreme poverty and become an incredible force for good. But all politics is local, goes the old phrase. And if you're in Pennsylvania or Birmingham you may not care that somebody in Delhi is getting a new job. Or that somebody in Jakarta is getting better health care. You care about what's happened to you and your family.
>
> And you're seeing this sense of stagnation, that you're not better off than your parents. And you're afraid that your kids may not be better off than you are. And what's the path forward? And the forces that are out there seem well beyond you. Globalization, immigration, trade, what else? You're afraid and you're looking for answers. And that fear, I think not just in the United States, but around the world, is what's fueling concerns, xenophobia, hatred, and a desire for answers that may or may not be there. It's a feeling of distrust of experts and disregard of traditional institutions. And we're trying to figure out how do we respond that that?
>
> What are the next steps for us before the world comes into this environment of tribalism that's self-destructive on the long term? There are these cycles of these things that often can last five to ten years before people feel as though, you know, they've had a chance to vent that anger. And yet we do think that history is on our side in a profound and important way that Martin Luther King made famous. A line that the moral arc of history is long, but

> it bends toward justice. I would say that the moral arc of history is long, but it bends toward progress. And out of progress comes rising living standards, healthcare, and ultimately the ability to transcend those forces of tribalism. Yes, reach towards justice.
>
> For five hundred years, technology and trade has risen and raised living standards around the world. And I think there's every sign that will continue. That is, we help that change come to pass. Well, it may be that the internet and globalization were part of the cause of this problem. We are also fundamentally an essential part of the solution to this problem.

The audience members were deathly silent, hanging onto his every word. It was a rhetorically powerful speech, but I saw it for what it was. It was the language of a cult leader, telling us who was good and who was evil. Maybe the Google employees were the problem, not understanding the legitimate concerns of the other side and addressing them as rational, decent human beings.

Walker continued:

> Prime Minister Matteo Renzi in Italy talks about the two worlds; the world of the wall and the world of the square. The world of the wall is a world of the fortress, a world of silo isolation and defensiveness. The world of the square, the piazza, the agora, the marketplace, where people come together in a community and enrich each other's lives. The tools that we build help people into the world of the square.
>
> You saw the video about the Missouri star quilt changing the fortunes of not just a family, not just a community, but the entire village was made better by the tools we make. Every day we help people come together, cooperate, communicate. Google is a trusted source of information for people around the world that's incredibly valuable at times like this.
>
> To make that happen, to figure out how we're going to navigate, not only continuing to make transformative products and making the world a better place. And yeah, I'll say it, even though they mock Silicon Valley for believing we need to be able to work together. We need to have each other's backs. We need to stand together in a time that's going to be incredibly difficult as we advocate for our values. And we see not only what the US administration, but the other administrations around the world take shape, how they shape over the next few years.
>
> I would say, please understand each other. Trust each other in the rule of law and let me turn it over to first Ruth, and then Eileen to talk about

> how we internally can continue that work of building bridges and working together.

Kent Walker appeared to view the Trump administration as a clear and present danger to Google's "values," preferring the world of "the square" to that of "the wall."

* * *

Ruth Porat, the chief financial officer of Google, was next on the podium. Ruth was a Stanford graduate, with an MBA from the University of Pennsylvania, Wharton School of Business, and a master of science from the London School of Economics. Prior to working at Google she'd been an executive vice president and chief financial officer of Morgan Stanley. She was in her mid-fifties, shoulder length brown hair, dressed more professionally than any of the three men who preceded her, wearing a dark blue blouse and slacks, as if she was den mother to a group of unruly, brilliant boys who might accidentally burn the house down if she didn't closely watch them. She began with complete candor:

> For what it's worth, I've been a very long time Hillary supporter. But as Ken said, the most important thing is I very much respect the outcome of the democratic process. And who any of us voted for is really not the point. Because the values that are held dear at this company transcend politics. Because we're going to constantly fight to preserve them.
>
> I want to take you back to 8:30 p.m. on Tuesday night. I was at home with friends and family watching the election returns. And as we started to see the direction of the voting, I reached out to someone close to me who was at the Javits Center where the big celebration was supposed to occur in New York City. Somebody who had been working on the campaign. I just sent him a note that said, you know, 'Are you okay? It looks like it's going the wrong way.' And I got back a very short, sad text that read, 'People are leaving. Staff is crying. We're going to lose.'

Porat stopped, bobbed from one side to the other, and then let out a deep breath of air. She shook her head, bobbed back to her original position, and bore down as if she wasn't sure she could continue. Tears welled up. Her voice started to break, and she brought her hands up as if to reach out to the audience. She continued:

> That was the first moment I felt like we were gonna lose. And it was massive, like a kick in the gut, that we were gonna lose. And it was really painful. And the thing that hit me, and we've talked about it before, like Sergey, my father was a refugee. And we moved to this country. And as a child I was always told, he fought hard, worked hard to get my sister and brother and I to this country. Because he wanted us to grow up in a place unlike what he had. A place where you would never be discriminated against based on who you were, the color of your skin, your religion, your beliefs. And that's the thing that kept going through my head on Tuesday.
>
> And it did feel like a ton of bricks dropped on my chest. And I've had a chance to talk to a lot of fellow Googlers and people have said different words similar to the concept of how painful it is. How painful this is. But I think there are three really important things that we should think about and talk about.
>
> First, throughout the campaign Hillary said we are a great country because we are a good country. And I firmly believe we are a good country.
>
> Second, one of the things that really struck me in her concession speech the next morning is she said, 'Please never stop believing that fighting for what's right is worth it.' And that is critical. We all have an obligation to fight for what's right and to never stop fighting for what's right. And that's one of the many things that I think makes this company so beautiful. Our values are strong. We will fight to protect them. And we will use the great strengths and resources and reach we have to continue to advance really important values.
>
> And the third message, that's super important, is the message from the election that a lot of people clearly felt disenfranchised, left out. We talked a lot about rising inequality. But how corrosive rising inequality is, is the other really important message from this. And on that we similarly have a very important role to play as do many others.

She paused for a moment.

> I think the main thing I just wanted to say is to give yourself time and space to deal with whatever you're going through. Healing is a process. It does take time. But one thing that makes Alphabet at Google so special, it's this term I've heard, I'd never heard it before I got here. Which is this is a place where you can bring your whole self to work. And we want everybody, wherever you were on the political spectrum, whatever it is, it's about respect for one another and continuing to ensure that we do that and making this a safe place where it's super clear everyone can bring their whole self to work and be

> respected. So, showing kindness to everyone around you is the most important thing. I feel super blessed to have had the opportunity to be a part of this community and especially at times like now.
>
> Yesterday Eileen and I had a town hall for some of our orgs and I suggested that what we all need right now is a hug. So, everybody, if you could turn around, or go to the person next to you, and do a hug, it works.

The camera on which I was watching the event from the YouTube headquarters panned to show the audience, many in their multi-colored beanies with a propeller on top, turning around in their seats and giving each other a hug. I found the entire event unbelievable.

Had there ever been such a nakedly political corporate reaction to an election?

Sure, certain industries were likely to be more or less in favor of a certain candidate or political party. But there had been something of an unwritten corporate rule in America that the workplace was free from politics. Google had not only crossed that line, in my opinion, but were acting as if no such line even existed. This wasn't just hidden bias, but right there in your face, for all the employees to see.

Ruth watched with approval the hugs being exchanged by the Google employees, then said, "Thank you, and with that I'm going to turn it over to Eileen."

* * *

Eileen Naughton walked to the stage and looked out at the group. She'd been at Google since 2007, working in media partnerships, leading sales and operations in the United Kingdom and Ireland, and leading their people operations team.[3] Although last to speak, Eileen was the most professionally dressed of the group, with short red hair, glasses perched casually atop her head, earrings, a gold necklace, white blouse, and dark blue jacket.

> All righty, so thank you, Ruth. Thanks for the hugs. I'm Eileen. I lead people operation at Google. I've been participating in TGIF for about ten years from remote offices in New York and London. And I kind of always imagined my first time up here. And people having good news around Google geist or something. But here we are. And you know, we talk a lot and we all know it. We talk a lot about what it is to be Googly. And I've seen so many instances and examples of Googliness in the last few days. The hug is one. I've seen

open and heartfelt communications. I've seen people feel safe sharing their thoughts, their dreams, their fears. I've seen Googlers show up for each other. Spontaneous groups of employee resource groups holding sessions, sharing notes, sharing resources, tips for how to get through hard times. I've seen gratitude and I've seen a lot of kindness these last few days.

So, let's try to internalize the kindness and keep it with us. I've seen Googlers talk about their differences from a place of tolerance and respect, and that's very heartening. And I've seen us try to intellectualize and understand the election results. Much as Googlers earlier this year when I was in London tried to understand the vote of the British people to exit the European Union. And just like with Brexit, I'm seeing people who are full of fear. They're full of fear about the future about what the uncertainty means for them and their families.

Since I'm in people, one of the questions I'm getting is how the Trump presidency might impact things like benefits and visas and jobs. So, there's a tremendous amount we don't know. I would just advise us all to be calm. You know there's a calm place that you can go to and just take a breath. It's obviously too soon to tell what the longer-term implications of the election will be. But we're watching closely, and, in the meantime, I thought I'd address three or four of the topics that we're hearing the most about.

First and foremost is immigration. We have nearly 10,000 Googlers in the US on visas. Very understandably those of you who are working here who have families here, or in the process of renewing or getting visas are probably very concerned. Here's what we know. There is for the time being, the Obama administration. There's no change, right? Googlers should not expect to be hassled entering the border. There should be no change in your status. We also know that the nature of the US immigration system is such that it makes, and there are legal limitations, it makes any immediate changes after January's inauguration of the new administration highly unlikely. But of course, we will keep a close watch on this. Our policy office in DC is all over it and we will keep you informed. But we will keep Googlers interests at heart. And we will, of course, fight to retain all the visas and then some, because we keep adding to this.

She continued, "The second question is around internal mobility. Can I move to Canada?" There was applause and laughter from the audience. I thought to myself, could anybody imagine a similar meeting at any American company, with the leadership talking publicly about employees wanting to move to Canada when Barack Obama won the presidency or re-election?

"The next thing we're hearing are concerns around our benefits and what does this mean, especially from Googlers who are concerned about benefits for their same sex partner. We are here behind you. We have led in this area. We will not in any way change our benefits." Applause from the audience and Eileen nodded in acknowledgment. "And we're very proud to take a very public stand on that issue, so nothing else changes."

> Then, finally, on diversity and inclusion. I think it's fairly obvious that Google leans largely liberal-democratic, but I do want to be clear that diversity also means diversity of opinion and political persuasion. And we value and welcome perspectives from all sides of the political spectrum. I have heard from some conservative Googlers in the past few days that they haven't felt entirely comfortable revealing who they are when these conversations come up at work. And so I believe we need to do better. We need to be tolerant inclusive, try to understand each other in this area. And you know, to emphasize that Sundar said in opening, and Sergey as well, you know the core values of civility, inclusion, respect, are what have always guided us and will continue to do so.
>
> I know you have a lot of questions. We're going to take them up. Here, we're also going to have some food and drinks on Charlie's patio for those who are here in Mountain View. I know there are large groups, probably Ann Arbor, Michigan, many offices in New York probably tuned in, so I hope you have something to drink. Water, helps, you know. You don't always have to add from one problem and start a whole other one. But anyway, with that, I think we'll take your questions.

The speakers had attempted to deal with the fact that roughly half of the country didn't agree with their choice, but it didn't come across as authentic. Yes, there were attempts to "reach out and understand." However, it came across in the tones of a principal telling the teachers not to be too harsh with a slow student. Did any of the conservatives, or even moderates, who might be able to understand both points of view, feel they could offer an opinion?

How can you offer an opinion when the co-founder, Sergey Brin, says he was "personally offended" by the election results? The chief legal officer, Kent Walker, talked about the election in terms of "xenophobia and hatred." And the chief financial officer, Ruth Porat, a person you'd expect to be logical and analytical, felt "we" had lost the election. How was any of it rational?

Did any of them understand how narcissistic and self-indulgent they sounded to at least half the country that might not have a few hundred million to several billion dollars in their bank accounts?

* * *

After the speeches, the leadership team of Google assembled to take questions via email and from employees present in the auditorium. Larry Page was a new addition to the group, not having given an actual speech. Larry was older, grey hair, wearing a bright yellow shirt, jeans, and tennis shoes, with a big smile that made him look like a mellow, aging hippie. He joined Sergey at the presentation on the right of the stage, while CEO Sundar Pichai, Chief Legal Officer Kent Walker, Chief Financial Officer Ruth Porat, and Head of Human Relations Eileen Naughton all took chairs in the center of the stage.

The first question, which came via email and was presented on a large screen, read: "This week's election demonstrated the problems with 'bubbles,' confirmation bias, and a failure to listen to other points of view. As technology becomes more personalized, how can Google reduce the impact of the 'echo chamber' and access to diverse viewpoints across the population?"

Sergey read out the question, then said:

> What should we say about that? I think that's a really big topic. Personally, I think there are several examples of bubbles. We saw a whole bunch of pollsters and so forth be really wrong. Not the pollsters themselves, but people interpreting the polls. And including probably all those folks getting ready to celebrate at the Javits Center, who I think didn't actually look at the data very carefully. So, yeah, there's definitely groupthink, that can be a huge risk. There's also just this story of two countries, you know, the divided nation and so forth.
>
> I personally think that's a little bit of an exaggeration of an explanation. When you actually look at all the data, there are definitely different folks in different walks of life and have different perspectives. And we definitely value seeing more points of view. There's also huge issues with trolling, state-sponsored trolling, you know, corporate motivated bias and so forth among media companies in multiple directions.
>
> Mind you, there's like a huge set of issues here all wrapped up in one question, and I don't think we can cover it all in one go. But I think maybe we should be more thoughtful about it. I mean, look, I think it's good, you know,

> it's a serious area. This is a deep issue. I think over times, as we, you know, are definitely in the role, the core mission to help users discover information, there are many, many places where we are ranking. We are algorithmically doing stuff, you know, so over time, understanding some of the things that are happening and course-correcting. I think it's good, but you know, it's a very, very difficult problem to tackle. But I think it's a moment of reflection, as I said earlier, for a lot of us.

In my brain, I tried to translate Sergey's rambling answer. When he said, "algorithmically doing stuff," I thought it covered a lot of ground. Here's what I thought might be the take-away. Yes, there are a lot of people with different ideas, and they don't often talk to each other, but with our algorithmic system of "ranking" things, we're going to provide the "approved" version of reality. I wasn't sure if that's what he meant, but it seemed to be heading in that direction.

* * *

It was near the end of the hour-long meeting, about fifty-four minutes in, when a question was asked which would irrevocably change my life. A young Google employee, somebody who looked like he could be a friend of mine or fellow team member, wearing a grey hoodie and glasses, with brown, fine hair, was called on by Sergey Brin.

The employee began:

> Google's mission is to organize the world's information and make it useful. But during this election we've seen a lot of misinformation, disinformation, fake news coming from fake news websites, being shared by millions of low information voters on social media. And ultimately there's been many, many people who've been voting, who've been acting, based on completely made up information. So, can Google do anything to try to figure this out? To try to do something against this very organized, very intense campaign of disinformation targeted at low information people?

The question was taken by Sundar Pichai, the CEO of Google: "Look, I think our investments in machine learning and AI [artificial intelligence] is a big opportunity here. You know, there is work we have done, the jigsaw team, did around what they call 'Conversation around AI,' to look at bullying, and commenting. And so, a lot of this is a problem of scale and not being able to keep up. Human systems fail and many of these things. So, I think investing in machine learning and AI could be one way we actually make progress on some of this stuff. But I think we should do more."

I sat at my desk at YouTube in San Bruno wondering if I'd heard the answer to that question correctly from my computer desktop. I understood how engineers thought, and what we say between the lines. Engineers are trained to identify problems and solve them.

But the meeting made it clear I didn't need to do much interpreting.

The election of Donald Trump was a PROBLEM, which needed a SOLUTION.

"Progress" needed to be made on that problem, and the name of the solution was "machine learning," which had already been developed. As Sundar Pichai had said, the real problem was how to "scale" it up.

Was Google planning to build a weapon against the newly elected president of the United States?

I was going to find out.

CHAPTER TWO

My Explosive Birth

BOOM!

On May 18, 1980, at 8:32 a.m., Mount Saint Helens in Washington State exploded. The blast released twenty-four megatons of thermal energy, equivalent to sixteen hundred times the amount of energy released by the atomic bomb dropped on Hiroshima at the end of World War II. More than a thousand feet of the north side of the mountain was obliterated. The base of the ash cloud generated was ten miles wide, at its mushroom shaped top forty miles wide, and fifteen miles high. A superheated mixture of gas, rock, and mud, called a pyroclastic flow, flattened vegetation and buildings over two hundred and thirty square miles. The crater from the explosion varied from between one to two miles wide, and half a mile deep. It was the biggest volcanic eruption in American history.

I'd come into the world less than two weeks earlier, on May 5, 1980, in sunny Southern California, at Hermosa Beach to be exact. My father worked for the local veteran's hospital in therapy and had a pending transfer to the Portland facility. Portland is located just a little more than fifty miles from Mount Saint Helens, so it had escaped significant destruction, but was blanketed by ash.

My mother and father wondered if the move was a good idea, but eventually agreed, despite the recent cataclysm. Our small family got to Oregon on June 21, entering a landscape where people were wearing dust caps and wiping ash off their cars in the morning. We lived first in Lake Oswego, a suburb of Portland, fed by the waters of the Willamette River, and then later in West Linn, a few miles away.

Obviously, I don't remember those days, but pictures of the time reveal a shattered landscape into which we had moved. Beautiful Lake Oswego was choked with ash, as was the mighty Willamette River, where my younger brother Trevor and I played on its recovering banks. As a child, I remembered the adults talking about the explosion. And I was often nervous when there would be rumblings from the mountain and minor explosions. I was assured there would never be another explosion as great as that of May 18, 1980.

One can never know how these experiences mold a person. However, I can't help but wonder if it made me believe, at some fundamental level, that the ground underneath our feet is never as stable as we think it is.

* * *

My brother Trevor was born on October 16, 1982 and that should have completed our family. But my mother and father had a contentious relationship and even the enormous energy needed to raise two young boys didn't help them mask those problems. The two separated when I was around five and my father moved to West Linn, a town just a few miles away. Trevor and I grew up commuting between their two households, trying to preserve a fragile peace between them.

My father described me as a deep thinker, even as a child. He'd say something while I was sitting at the kitchen table, I'd cock my head, and stare off into space, as if trying to work though some complex thinking. I still have that behavior, so if you ask me a challenging question, don't expect a quick, glib answer. Sometimes that silence can unnerve people.

Although that trait has been useful as an adult, it caused problems for me as a child. I went to kindergarten at a Christian school and the teacher met with my parents to tell them she didn't think I was okay. I was often distracted in class, with my mind somewhere else. She tried to characterize me with some terminology that was popular in the day. But to their credit, my parents didn't buy it. They pulled me out and put me in kindergarten at the local public school.

Repeating kindergarten was probably one of the best things my parents ever did for me. In addition, it put me closer in school to my brother, Trevor. That would end up being very important as in high school our social circles would eventually merge.

Many of my best friends today were originally Trevor's friends.

My father says I was inquisitive, peppering him with so many questions about everything that he often found it fatiguing. When we'd go on family trips, he instituted a policy that I had a quota of fifty questions I could ask during the drive. Usually I'd hit that mark at about an hour, and I'd sit there fuming for the rest of the trip. We never reached a resolution on whether a question he couldn't answer counted toward the fifty questions. I maintained that a question he couldn't answer shouldn't be included in my quota.

* * *

As kids, Trevor and I couldn't have been more different.

I was an introverted intellectual while my brother was rough and tumble, athletic, popular, and the center of attention in any room he entered. He was aggressive, and when his temper rose, he got this look on his face that we knew meant he was going into "beast mode." We'd watch him in some sporting event, see that look, turn to each other, and say, "Uh-oh, Trevor's going into "beast mode."

Although we got into it a few times while growing up, by our teen years he was bigger than me and could kick my ass, so I didn't have much interest in physical conflict with him. Trevor excelled at all the sports he tried, including soccer, football, and wrestling. My father vainly encouraged me to pursue sports, but I struggled to hold attention for it. The only success I can report is I became a moderately good soccer player.

Since I was just a grade ahead of Trevor, it was easy for our social groups to begin to overlap, especially as we entered high school. He created a tight group of friends, similar to him.

At one point, Trevor became a cage fighter and fought twice in something called "Desert Brawl," which took place in a giant octagon. A friend of mine was working for the operation and developed an online ticketing system for the event I programmed. We also created an instant replay system for the crowd by rigging up a TIVO system that would loop back into the video system as well as a "streaming" system for the cage fights so people could watch the event over the internet, which was very advanced for its time in the early 2000s.

Trevor's first fight in the Octagon didn't start well. He was fighting a guy named K.C. Jones, an experienced fighter who was no stranger to the cage. Trevor was getting owned by the fighter but managed to get a hold of the guy's arm and NOT LET GO. (I think this is a Vorhies family trait.) The guy used the grab to pick Trevor up and drop him on his head. But Trevor never lost his grip. Trevor used his whole body to position the guy's arm into just the right position to apply an "arm bar." This move lets the opponent know you can snap his elbow in the wrong direction if he resists. The guy tapped out. The fight was over.

Trevor was so dazed that he started exiting the octagon before the referee could declare a winner. The referee had to race over to Trevor, take his arm, raise it up, and declare him the winner, to the cheers of the crowd.

As soon as Trevor got out of the cage, he said, "Get me a bucket." One was quickly handed to him and he immediately puked into it.

The second fighter Trevor went up against was a guy I introduced myself to before the fight. He was an impressive physical specimen, muscles bulging out everywhere, and a few inches taller than me. In the course of our conversation he mentioned he had a shoulder injury.

That fight took place in a bigger venue than the first fight, a large stadium filled with people. To heighten the drama, the main lights were turned off, and the cage was lit up like a movie set. The crowd was rowdy and ready for their blood sport.

As Trevor was about to enter the ring, I approached him and said "Hey, you won't believe this, but I was just talking to your opponent. He asked if you could take it easy on him because he's got a lame shoulder." I paused for a beat and then said, "Do you understand what I mean?"

Trevor got a little, wicked smile on his face and nodded.

Because I fulfilled the exact request of his opponent, I feel no guilt over what happened next. Trevor got that "beast mode" look in his eyes and fixated on his opponent's shoulder like some great predator about to pounce. I don't think he even blinked in those few seconds before the bell rang.

When the fight started, Trevor charged his opponent like a lion on the hunt. With grace and beauty, he quickly brought his opponent down, pressing toward his injured shoulder. The guy immediately tapped out. The whole fight didn't last more than ten seconds. The opponent's shoulder was fine. Trevor had gone "easy" on it. But he made it clear he'd go farther if needed.

In his first fight, Trevor hadn't been paid by the organizers, and the same thing happened in his second match. Trevor considered that a sign, and quit cage-fighting, with a 2–0 record.

In his professional life, Trevor became a medical sales representative, which is a perfect fit for his outgoing personality. Some people want others in their life to be exactly like them, in interests, politics, or worldview. But I've found I take away the most from people who are different than me.

Trevor was lucky to find and marry a wonderful woman, Hannah. She's warm, nurturing, open-minded, and just has this wonderful curiosity about the world I find refreshing. I've observed her at some pretty terrifying, and in retrospect, hilarious moments. Their wedding day is a perfect example.

Trevor had rented himself a beautiful Swiss-style chalet on a gorge overlooking the Columbia River, near Mount Hood. They decided to have the wedding there, setting up tents, chairs, and a stage for the band a couple hundred feet away. It was a beautiful location, surrounded by forests, and couldn't have been more idyllic, with stunning views of the Columbia River. They exchanged vows, we partied, and eventually I passed out, sleeping in a small tent I'd pitched in a quiet section of the woods, knowing I didn't want to drive afterwards.

About thirty minutes later one of my friends, William, kept calling my phone and finally I answered, mumbling a greeting.

"Dude, there's a fire! You need to get out!" he said.

"What?" I replied. I heard the distant crackling and popping of wood and saw a faint glow through the tent wall. I scrambled and found my glasses.

I quickly opened the flap of my tent and saw a big glow emanating from the direction of Trevor's house. I saw little floating ash embers in the air. A fire had started down at the bottom of a cliff below Trevor's rented house. I raced to the house to see it engulfed in flames.

"I can't find Trevor!" screamed Hannah. "I think he's in the house!"

A family member dashed into the house, ablaze and filled with smoke, checking all the rooms, but only finding Trevor's dog, a small pug that was very relieved to be rescued. At about that time, Trevor walked up in his tuxedo.

The house was a complete loss, but thankfully, nobody was hurt.

I put together a Go-Fund Me account to help Trevor recover some of the money and raised nearly seven thousand dollars.

* * *

One of the things you should know is I have a very small family. Neither of my parents had siblings, which means we grew up without aunts, uncles, or first cousins. The only grandparents I knew were on my dad's side. My mother's parents died when she was in her twenties, within six months of each other. That meant for all practical purposes our family numbered a whopping six people. My father's parents died a few years ago, so when that happened it knocked us down to four.

I wasn't a good student, finding it hard to pay attention, and teachers were usually writing letters home to my dad about me having focus problems. From about the age of ten I've also suffered from debilitating migraines, usually triggered by eating anything with grains. It's likely I've got at least a touch of Asperger's Syndrome, as some human behaviors that others take for granted seem a little alien and strange to me.

Up until fifth grade I had a succession of teachers who I either liked because they were kind to me or exasperated by my lack of attention. But in fifth grade there was a district split and I was sent to Willamette Primary School, where I fell deeply in love with computers. You might as well say the heavens opened and the angels started singing for the impact that would have on my life. We started the year with an older teacher, but she was then replaced by a young student teacher. And that's when I was introduced to my first great love, the Macintosh Classic.

I'd worked with an Apple II GS in third grade, but it was clunky and difficult to get into. I didn't love computers at that point but was able to play various games like Marble Madness, Mixed Up Mother Goose, and Oregon Trail.

But the Macintosh Classic I encountered in fifth grade created TOTAL COMPUTER ADDICTION in me. I recall being mesmerized by even the screensavers which showed flying toasters, or a fish tank where if a small fish swam too close to a big fish it would be eaten. The computer had hyper-card games (an early software application for the Macintosh that was used for games), and I'd be so fascinated that I'd often work on it in the classroom during recess, rather than go out and play in the gloomy Portland weather.

But then in the following year, in sixth grade, my mom surprised me with a Macintosh LC II at Christmas. It was like a souped up Macintosh Classic, BUT with 256 colors!! Quite an achievement at that time and very expensive, at about three thousand dollars It was a monumental gift, considering that my mom had little income at the time and bought the computer on credit.

By sixth grade, Macintosh computers were everywhere, with multiple computers in each school room. In sixth grade, I learned how to type on the computer and was suddenly so much quicker that I'd been with a pencil. I also learned about transferring files from one computer to another and the terrifying concept of computer viruses. Other kids were also intrigued by the computer, and I loved teaching them all I learned.

By seventh grade I was able to make my own computer virus, altering the hyper-card virus called "the merry X-mas virus" with a hex-editor. Of course, after I tested it on a computer to make sure it worked, I wiped it clean. I didn't want my monster getting loose in the world. I might have been a mad computer scientist, but I was a responsible one.

* * *

When I got into high school, I made a decision which completely changed my life.

I tried out for the speech and debate club. In the club, you had to defend yourself from really smart people who would demolish your arguments in the way of a good comedian, often with withering sarcasm.

At first, I was terrible, getting ripped apart by that class of wolves. And yet it was as addicting as any drug to me. There was an undeniable anxiety as I got ready to present. But even if my arguments got dismantled, I was still buzzing with the rush of it. I knew I had to make the club a central part of my life.

However, I was terrible in those early days. The speech and debate club was attended by theater kids with a dark sense of humor. I found myself laughing constantly at the comedy gold coming out of their mouths. The club often had impromptu debates without notes, dissing each other with facts and put-downs, making it the closest thing West Linn ever had to a rap battle. I was so intimidated by the eclectic cast of characters that I simply watched quietly for weeks.

My baptism of fire was when they invited me to go to a speech and debate tournament. I signed up for an impromptu debate event. The club leader, Adrienne, a hippie girl who always wore purple fuzzy clothing, accompanied me to the event for moral support. I was so nervous I froze up for the first thirty seconds I was supposed to speak.

But finally, glacially, words started coming out of my mouth and I was able to produce some sort of impromptu speech about the topic.

As I walked out with Adrienne, I told her the impromptu speech was a total disaster. But after each subsequent foray onto the intellectual battle-field of debate my anxiety would decrease. Impromptu speaking did some-thing weird to my cognition, causing me to become completely focused, synchronizing my entire brain to concentrate on that single task. Besides riding motorcycles, speech and debate competitions are the only thing in my life that has given me such complete Zen-like concentration.

Even though I was still new at it, I quickly realized the high anxiety was good for my performance. In practice, I was quick to volunteer for impromptu sessions and by my second competition, my brain was firing on all cylinders.

By the end of the year I was doing so well with impromptu speaking that I made it into the regional finals, taking third place as a freshman. Adrienne was shocked when she saw my name on the final list. At the end of the year dinner she gave me an award entitled, "Well, That Happened Fast!" to the laughter and applause of my teammates.

I went from being the worst impromptu speaker in class to having some real talent, regularly competing in regional finals. I transitioned out of playing sports to year-round speech and debate. Among my fellow speech and debate kids, I felt I'd found my tribe. The kids were super-intellectual and into the nerdy things I liked. In addition to our weekly meetings, we traveled around on the weekends to speech and debate competitions.

Speech and debate also led me to the University of Oregon, and specifically the Eugene campus. One of the regional competitions was held over a weekend, Friday, Saturday, and Sunday, at the Eugene campus.

I remember walking around that beautiful campus and falling in love with it, imagining my future self as an enrolled student.

* * *

When it came time to apply for college, there was only one place that appealed to me from my years in speech and debate.

In retrospect, the University of Oregon at Eugene was a terrible choice, especially for my first two years. The University was a liberal arts school which meant it wouldn't be until my junior year that I got to take an actual programming class. While on paper I was in an engineering track, their idea of a computer class was a mathematics class with a chalkboard and chalk, discussing and working on algorithms with pencil and paper, the way they might have done in the 1960s.

But eventually in my junior year I got into a class with actual computers and was ready to roll. I loved programming so much that I'd start that night on the assignment, often to the detriment of my other classes, where I'd do the homework at the last minute.

I was lucky enough to get a job at the university, monitoring their computer network from a special room. I was supposed to insert a back-up tape on a certain schedule so the main university computer disks could be copied to tape archives. In addition, if anything happened, I was to call a supervisor, tell him what had happened, and then insert another tape into the computer. The job left me with a lot of free time, which I was able to use to do my homework. The pay was also good, which meant when I left college, I had very little student debt.

I'd long dreamed of becoming a video-game developer. By happy coincidence, there were some video-game development companies in Eugene. The best one was named Pipeworks and when I was in college, I landed an interview with them.

They brought me into an interview room. After some initial conversations, the interviewer drew a challenge on the whiteboard called the "triangle program" and asked me to solve it.

I couldn't do it.

The interviewer said, "This involves really basic math you need to understand if you want to work at a video game company." I left feeling humiliated.

That interview changed my college plans.

I'd been planning to major only in Computer Science, but it was clear I was missing even the fundamental mathematics I'd need for creating computer graphics. To solve that, I decided to major in mathematics, as well as computer sciences and psychology. I imagined some future potential employer looking at my resume and somebody else who'd recently graduated from college. My competition would have one, possibly two majors, but never a third. I figured a third major would make me stand out in any future job interview.

* * *

In college, I had a good friend named Graham. We'd gone to high school together and been friends, but he'd been a year ahead of me. We became better friends in college. We always had excellent conversations, talking freely about any subject that crossed our minds.

In addition, he was often coming up with interesting ideas, and I was usually quick to jump on board. Having three majors required me to spend a few extra years in college. But since I had a well-paying job on campus, shouldn't I explore other facets of life?

Graham decided he wanted to get a motorcycle license. Of course, I had to get one as well. While Graham got himself a nice bike and the best gear, I got an old 1989 bike, but with practically no miles on it. It was cheap and slightly purple, but if I wore blue biking gear, I looked good. My love life dramatically improved when I became Zach, the motorcycle guy.

After a few months, Graham joined a motorcycle riding group. I did the same shortly afterward. The club liked to ride long distances, and we'd go on roads around the outskirts of Eugene, Oregon. It was amazing to go from the cityscape of Eugene to the rolling hills covered with pine trees and beautiful valleys with meandering rivers.

The riding group had one drawback. They liked to go fast. Really fast. One time we were on a long, open stretch of road and I was going a hundred and thirty-five miles an hour when another rider passed me like I was in first gear.

My first thought was, "Man, I need a faster bike." But right after that I said to myself,

"No, I don't. I need to stop riding with these people. This is dangerous."

About an hour later I was on that same road, heading back to Eugene, when there was a bend in the road. I started to slow down, but not enough.

I started to drift toward the gravel shoulder. Beyond the shoulder was a ditch. If I hit the loose gravel I was going into the ditch. I was torqueing my bike as much as possible, knowing I didn't want to slap my knee on the ground going a hundred miles an hour. That might make me flip and tumble on the asphalt road.

I focused, sending up a prayer to God to save this stupid kid, as I watched my front wheel come within an inch of the gravel. I felt the force of gravity shift from sideways to forward, tires gaining full traction on the road, and I rocketed away from danger. I slowed the bike down to a casual eighty miles an hour and took several deep breaths. I really needed to think about leaving the riding group.

My decision to leave the biking group was sealed when I saw a terrible accident, caused by exactly the kind of showboating I'd be likely to do if I had the talent.

There was a guy who had an awesome sports bike, a Yamaha YZF-R1, I think, and he could do amazing tricks on it. One of his tricks was to ride on the back wheel while still steering the bike. It seemed so exhilarating and I longed to try it.

That wasn't enough for the trickster, though. He perfected a stunt where he'd ride the motorcycle in a wheelie, but with him sitting on the bike in the opposite direction. Yes, that's right, his back was on the bike, and he steered by looking behind him.

But that time, unfortunately, he steered himself into a curb.

CRASH!

The bike launched itself into the air, with him along for the ride. The bike tumbled end over end. Now, a human body is flexible and will quickly slow down when it encounters the ground, bending at all the necessary places. But this guy's leg was still wrapped up with the motorcycle. The bike kept going, but his leg snapped. Everyone rushed to see what happened. His ankle was pointed at his head. It nearly made me sick. He wouldn't be doing those tricks again for months, years, maybe even never, if he was smart enough.

That was my cue to exit the scene and drive home, obeying every traffic law on the way. I considered it my message from the universe to give up on the whole biker gang thing, using my motorcycle only for transportation and occasional vacation rides at reasonable speeds.

* * *

As luck would have it, I ended up interviewing with Pipeworks again in my final year of college.

My interview was with the SAME person and he gave me the EXACT same test I'd failed disastrously years earlier. But that time I was able to solve the problem using trigonometry.

It was a good solution . . . but not the one they wanted.

However, nobody had ever solved the problem using trigonometry before. (They wanted vector subtraction, a way I hadn't even considered.) My solution was super simple if one thought about it from the standpoint of trigonometry. But up to that point, nobody at the company, or who'd interviewed with them, had used that approach. The interviewer looked at it curiously for a moment, then smiled, and said, "You know what? I really like it when someone comes up with a non-standard solution. Because then I get to see how they think."

I was hired on the spot and reported to work the following Monday.

I waited several months to tell my new boss he'd interviewed me a few years earlier, and it had been a complete failure. I imagined it was such a deep, dark, shameful secret.

But when I told him he simply shrugged and said, "Oh, that's funny."

Pipeworks had been founded in November 1999. Their first game was a Tetris-like shape-matching game called *GLOM*. They struck it big with their next title, *Godzilla: Destroy All Monsters Melee*, then *Godzilla: Save the Earth*. That was followed by *Prince of Persia: Revelations*. The project I was assigned to was *Rampage: Total Destruction*, the fourth game in the *Rampage* series.

The first *Rampage* was released as a three-person arcade game in 1986. It quickly became one of the most popular games of the year. Players had a choice of being one of three monsters. There was George, a gigantic King Kong-like gorilla, Lizzie, a half dinosaur/half lizard creature, and Ralph, an enormous werewolf.

Gameplay was simple and satisfying, consisting of destroying buildings by punching them, as well as any vehicles you might encounter. If you were damaged by any enemy bullets, falls from buildings, punches from other monsters, shells from tanks, or blasts from sticks of dynamite, you could regain energy by eating various items in the game like roast chicken, fruit, or even the occasional fleeing soldier. You moved onto the next level by destroying all the buildings in the city.

By the time I started working on the game in 2005, there'd already been two other versions of the game. How could one continue to keep it interesting? I was made the game programmer on the game.

I thought the key to success would be in the details. I made it so the human characters would fly toward the camera after the monsters kicked them. Punching trains, trolleys, and police cars in the game was a constant from the first release. But the vehicles didn't crumple in a realistic manner. I made it so when one of the monsters hit a vehicle it would crumple from the point of impact and travel like a wave through it. I added smoke and fire effects to the train collisions and tried to make the devastation caused by the giant monsters more closely resemble what you might see in a big budget Hollywood movie.

We made a presentation to our client, Midway Games, and they absolutely loved the changes we'd designed. I received high praise for the work and my salary was increased by a third, which was great as I was still technically a college student. I was making serious money. All of that happened within six weeks of joining the company.

The future looked bright for young Zach.

CHAPTER THREE

I "Chose Poorly" with Indiana Jones, but Still Saved Earth

I can be argumentative and opinionated. These qualities have had some consequences in my life.

But when I've been bitten in the ass for speaking my mind, I've generally failed upward. I'd describe myself as an "anti-collectivist." If my boss suggests a plan of action I think is wrong, I'll usually resist the temptation to go along and might even say, "This plan is stupid. I have a better idea."

It takes a certain kind of person to manage me, and I understand that. I need to respect their intelligence and experience. And they need to know I can have some sharp edges, but I'm not trying to be rude if I cast doubt on an idea. As I get older, I try to change as well. Learn to disagree in a way that's, well, not so, disagreeable.

When I worked at Pipeworks on the *Rampage* project, I had a boss I needed to work around. I didn't like his ideas about the scripting engine. But I was able to work quietly on what I thought would succeed, then when the presentation was made to the client, they loved what I'd done. I realized my social skills weren't the best, but I came through in the clutch.

I couldn't do that with my second boss. He was arrogant, very smart, and kind of a jerk.

He put me on a difficult problem, creating a system for a racing game which tells players how to build a faster car. That was out of my league as an entry level developer. He just threw me into it, didn't offer much help when I asked questions, and then when I couldn't deliver, said, "Zach, I don't think it's working out. I think you need to go work somewhere else."

After getting fired, I put my resume on a website called *Gama-Sutra*, a play on *Kama-Sutra*, the ancient Indian text on love-making (who says computer nerds don't have a sense of humor?). I immediately got a response from Lucas Arts, located in the Presidio section of San Francisco, right near the Golden Gate Bridge. (Lucas Arts is a division of Lucas Film, the production company put together by director George Lucas.)

I talked to the recruiter over the phone but didn't make a good enough impression for them to want to continue the hiring process. A few weeks later, I decided to go down to the San Francisco Bay Area to interview at a different company. I let Lucas Arts know I was going to be in town in case they wanted me to interview me in person. To my delight, they said sure.

I went for the interview, it went great, and that night they called to offer me a job. And I'd be making more than double what I'd been making at Pipeworks.

One of my big problems was finding a place to live. Even though Eugene was a city, it had a small-town feel. If you were interested in an apartment, you'd talk to the owner, see if you liked the place; if the owner felt comfortable with you, and if everything else lined up, you'd give him a couple references. In San Francisco in 2007 it was completely different. For just a small room in a house there'd be twenty people lined up with completed rental applications and their list of references.

But I needed a place to live and I needed it fast.

I found a beautiful hostel in Fort Mason at the edge of the Marina District. Although it was like living in a military barracks, with eight people to a room and sleeping on bunk beds, it was surrounded by trees and right on the ocean. I'd get up in the morning and eat waffles covered with butter and syrup while looking out at Alcatraz and sailboats cruising on San Francisco Bay. Then I'd hop on my motorcycle and cruise through the streets of San Francisco to new my job at Lucas Arts as a gameplay engineer.

When I started at Lucas Arts, they happened to be in transition. They'd been working on two projects, a Star Wars game and an Indiana Jones game. Progress on both projects had been poor. The entire game company of Lucas Arts had been fired, except for a few employees. The Star Wars team had been rebuilt and they were about halfway through the game development cycle. The Indiana Jones team had just been started, and they were still in a preproduction stage.

Although there was still some room left on the Star Wars team, there were more slots available in the Indiana Jones group. I wanted to be on a growing team and be one of the foundational members. So I chose to be on the Indiana Jones team. As the immortal Grail knight says in *Indiana Jones and the Last Crusade*, after watching one of the villains chose the wrong Grail cup and decompose in front of the audience's eyes—I "chose poorly."

I should have picked the Star Wars team.

At first there wasn't much to do because they were assembling the new team, so they gave me busy work. Things like "look at this new physics engine and write some demos for it."

I'd started working on a few tech demos, then they wanted me to start working on the AI (artificial intelligence) of the game. We came up with the idea of "smart objects," meaning when you picked them up, the game engine instructed what to do with them. For example, let's say Indy picked up a pick axe. It would contain all the information for the character AI to be able to use it.

As the team grew, we got a new boss, and that's when I started having trouble. I figured the best way around the guy was to put in an insane number of hours, produce superior work, and get high marks, as I'd done when I first joined Pipeworks. My early work on the AI received high marks. I figured I could "push through" with my ideas by putting in extra work and showing they were superior. I hoped if the stakeholders were impressed, they'd say "Zach did a fine job with the characters interacting with Indiana Jones. Let's trust Zach to see out his vision." The final demo our group came up with actually scored high in the AI department, and was in fact the highest rated part of the game.

I felt vindicated, only to be brought into the office and told I was being put on a "performance improvement plan."

It seemed I worked well with machines, but not so great with human beings.

The "performance improvement plan" required me to meet with the human resources team once a week. In the first meeting, they were nice, telling me I was going to make it and saying we simply needed to straighten things out. At the second meeting, I thought they'd acknowledge my improvement as I was trying to be more collaborative and less confrontational.

But they slammed me.

Throughout the next week my boss gave me several hints they were going to get rid of me. The night before my final interview I packed up all my things.

As expected, in my third interview with human resources they said they were terminating me. In ten months, I'd landed and lost my dream job.

They gave me a nice severance package though, with a couple months' salary.

It was October in San Francisco, which is kind of like their summer, with warm, calm days. I remember sitting on the sand dunes at Ocean Beach, looking out at the water and thinking to myself, *This really isn't so bad. This might just be fantastic. I learned a lot. And I got as far as I did because of what's in my head. And that's all still there.*

* * *

I started to apply for jobs and one of the places that responded quickly was Google.

I did an interview over the phone and nailed it. An hour was scheduled, but I finished in forty-five minutes and the interviewer said, "Well, we're done, but let's keep talking." Later I learned I'd gotten an essentially perfect score on the phone interview.

They brought me in for an interview and then gave me problems to solve on the white board. Not only did I solve the problems they gave me, but I also nailed the harder, more optimized versions of the problems.

At the time, in late 2007, early 2008, it normally took Google three to four months to take a person from first interview to sitting at their desk at the Google headquarters in Mountain View. I was hired in just under six weeks.

One of the things you should know is that Google didn't hire people for a specific position. They liked to hire smart people and then figure out where you might best fit into the company. Because of my experience writing artificial intelligence (AI), as well as graphics, I was selected to interview with the Google Earth team. At the time, they were trying to pair their Google Earth program with the Audi navigation system.

I thought it was a fantastic idea. In addition, my manager would be the guy I interviewed with over the phone. I'd liked him from the start. And the more I got to know him, the greater my respect for him became. He had an encyclopedic knowledge on how to do things and was quick at providing answers. I'd give him a problem, and within a minute or so he would have sketched out a possible solution and handed it back to me.

I'd look at what he did and usually say, "Oh, wow! That problem really was possible to solve!"

When I started working on the Google Earth application for the Audi project, I quickly identified the main problem. The embedded system they were using was significantly underpowered. A typical laptop you'd use today is measured in gigahertz. The Audi system was measured at 600 mega-hertz. Mega in the computer world means a million bytes, while giga means a billion bytes. Get the difference? It was like we were using 1992 computing capacity in 2008.

We had a six hundred-megahertz processor and because of the way the system was configured we could only get a third of the processing power to focus on creating the roads. In addition, each time a new frame was created, the system had to completely recreate the entire road system. It was an enormous use of computing resources for just the roads.

I brought the problem to my manager, but he didn't want to fix it. I pressed the issue a couple more times, but he became even more vocal about not addressing it. I was about to go into old Zach mode where I'd simply ignore his instructions and continue, when I stopped myself. I thought, *Don't mess this up, Zach. How can you fail upward from here?*

I bit my tongue and abandoned my solution to the problem.

About two months later my manager came to my desk and sat down. "This project is going to fail," he said. "The processing power isn't enough to run at twelve frames per second. Unless you have a trick up your sleeve, we're going to lose this contract with Audi."

"As a matter of fact, I do," I replied. I explained the problem was the need to recreate the roads with every new frame. The geometry was unstable. With a little modification, the roads would remain stable and skip recomputation.

My manager perked up a little, still skeptical. "Prove it," he said.

"I'll get to work on it right now," I told him.

The next day I had a demo. Two weeks later I had a working prototype. And six weeks later, I had the entire project done and ready for review. The speed improvement was massive. By moving the system to a persistent geometry, we cut the amount of data needed from 66 percent to 1 percent. The system was now back into its original technical parameters.

That trick up my sleeve saved the Audi project. And not only did they put the system into their A8, but quickly added it to their A6. The project would eventually make Google hundreds of millions of dollars. The deal gave a surprise boost in the financial statements of the entire department with a mention in the company wide financial statements during one of the quarters in 2012. I was able to put that accomplishment on my paperwork for my next review cycle.

It was personally important to me because I was trying to reach the position of "senior software engineer" at Google. Normally it takes four years to reach that level. In making my application I solicited six people for advice. Half of them said I was one of the smarter members of the team, and the other half said I was challenging to work with.

Sensing a theme in my life?

Even though half of the people I consulted thought I wasn't ready, I applied for the position after about three years into my time at Google. To the surprise of many, I got the promotion. That gave me additional status, and honestly, I didn't want to go any higher. There were positions above senior software engineer, such as principal engineer. But I didn't want to be a principal engineer because the workload was insane, or staff engineer, which meant I'd be supervising the work of others. I'm much better with machines than people, and I appreciate that. As Clint Eastwood once famously said, "A man's got to know his limitations."

* * *

While my work was important, I never wanted to be a drone who lived only for my job.

One of the common rites of passage if you were working in the tech field during the late 2000s was the Burning Man Festival in the Nevada desert. For those who've never been to the festival, it's a unique experience and requires some description.

Burning Man is best described as an unlikely party in the desert. The physical location is inhospitable, a dry lakebed with alkaline, caustic soil. A temporary city is constructed at the location which lives for a week, like some beautiful, complex, short-lived insect, then on Sunday it's all torn down. To mark the end of the temporary civilization, on Saturday night a large wooden figure is burned as darkness falls and then on Sunday the temple is burned.

The festival officially opens on a Monday. All the buildings and artwork appear shiny and new, then as the week progresses everything falls into disrepair, until it all gets deconstructed on Sunday. At Burning Man, one gets a sense of how societies rise and fall. Those who show up on Monday are the true believers who want to be swept up in the experience, and have lots of intentions to better themselves or explore different parts of their personality.

Those who show up on Thursday or Friday are often just interested in partying for a few days, raging like barbarians, and heading home hungover on Monday.

But first I had to get to Burning Man. I didn't want to ride my motorcycle all the way to Nevada. And besides, when you go to Burning Man, you need to provide your own accommodations, usually a tent, which I had.

I saw an ad in a local paper offering a ride in an RV, if I helped pay for gas. It seemed like a good deal, and the guy (I'll call him "John") lived in San Rafael, just across the Golden Gate Bridge in Marin County. I called the number and talked to a woman named Malia, who had this unique, sing-song way of talking that I found delightful. She explained she was a tenant and providing some help to John as he'd recently gone through some difficulties. Because of Malia I figured things were fine and made plans to go to their house on the day they were planning to leave.

I drove my motorcycle to the address given to me and found myself looking at this gigantic, opulent house. But when you looked closely you realized it was in poor shape, as if nobody had done much upkeep in the last ten years. I rang the doorbell and Malia answered. She had this fiery red hair and green eyes that were just so full of life. We talked for a minute or two, and then John, an older man, entered the room.

After a brief introduction, John said, "Hey, I want to show you something," and motioned for me to follow. He took me down to the basement and opened a door. I had to shield my eyes from this brilliant orange light emanating from the room. When my eyes adjusted, I saw hundreds of pot plants in the room he was using as a subterranean grow facility. I made some appreciative comments, not knowing what to say; he seemed pleased, and we went back upstairs.

I should've seen more of it when I first arrived at the house—the corners of rooms filled with junk, the riot of dogs living there, and the sense the place had once been opulent, but was falling into ruin. Despite the strangeness, he seemed to have a lot of people around him, almost like a cult leader.

I thought to myself, *This might be a mistake. And a total shit show. But at least I won't be bored*. I decided to throw in with that rag-tag group of people going to Burning Man.

One of the things you do at Burning Man is you give yourself a new name, known as your Playa name, like you're at the beach. One of the guys who joined us was named Jason, but he chose "Trash" as his Playa name. It felt weird to call somebody Trash, but he was a pretty cool kid. My Playa name was "Strider," because I had bought some running shoes that were springy and bouncy.

When I stumbled out of the RV at the amazing city which had sprung up on that dry lakebed in Nevada, I felt something akin to a spiritual experience. I'd never seen so much art in a single place and it made me realize the drabness of my regular life. On that first day, I made a resolution to surround myself with more art.

When night fell, I realized the critical mistake I'd made. I brought a tent, but it wasn't air-tight against the dust. My tent had these vents, with the assumption if you were out in the wilderness you wanted that fresh air. The assumption was you'd likely be using the tent in some mountain pine forest, not a dry lakebed in a desert. When the wind started to blow, it picked up that caustic lip-cracking dust and pushed it into my tent. One night there was a dust storm, and I woke up covered with dust, my lips chapped, and my eyes watering. Yes, it was uncomfortable, but hey, it was Burning Man. The next time, I'd plan better.

* * *

Malia was originally from New York, and her father died too young and left her an enormous apartment at 65th and Columbus, right next to a major rail line. The building was more than a hundred years old and had a long hallway with a full-length mirror, the kind you'd find in a dance studio. I was at Malia's to take part in the Occupy Wall Street movement in New York.

Malia's apartment wasn't far from Zuccotti Park, the epicenter of the Occupy Wall Street protests. I was starting to become political at that point, wanting to contribute my voice as a citizen. Since I could just as easily work out of the New York office as the Mountain View office, I requested a transfer of a few weeks so I could go to the protests.

The request was granted, and Malia let me stay at her apartment. During the day, I was working at Google, but at night I went to the protests, bringing food and supplies to those protestors who were camping out. In addition, I had an iPhone movie projector and started showing documentaries at night about the corruption of corporations and the nature of the Federal Reserve. It kind of became my shtick. I was Zach, the movie guy, providing the evening's entertainment. The protests really appealed to me, maybe because I have something of a counter-culturist bent. I was genuinely sympathetic to the idea of people coming together in one voice and saying,

"We're mad as hell, and we're not going to take it anymore!" At one point, I even spent a couple nights sleeping under the tarps set up by the protestors.

As the protests continued, I returned a few more times to New York. But as they grew across the country, I found that protests were springing up in California, as well.

* * *

In November 2012, there was an Occupy Oakland event, on Telegraph Avenue, and I decided to go. There'd been reports of violence breaking out at the rallies, with the blame being placed on overzealous law enforcement. That was the time when citizen journalists like Tim Pool were doing underground reporting by using simple Go-Pro cameras to document events. I wanted to follow in their footsteps. I imagined myself breaking a national story about police brutality against the peaceful Occupy Wall Street community I'd come to know and love.

In addition to my Go-Pro camera, I took my motorcycle helmet, as well as my motorcycle body armor, built with Kevlar, the same material used in bullet-proof vests. The motorcycle body armor was designed to help those of us who might go a little too fast on our bikes better survive a high-speed crash.

But there was a new group at that protest.

They called themselves "anti-capitalists" and were the forerunner to Antifa.

They were dressed in black clothes, carrying backpacks, and black flags with thick, wooden poles. At what seemed to be a pre-determined signal, they all dropped their backpacks, slipped on black clothes and facemasks, and started to use their thick, wooden flag poles to break the windows of local stores.

I was stunned by the scene. They were not the peaceful, but noisy Occupy Wall Street protestors I'd known for years. Their actions were dark, malevolent, and preplanned. I'd been carrying my body armor to the protest, so I decided it was time to suit up.

I put my motorcycle helmet on as well, and as we used to say about my brother Trevor, I was going "hardcore." It was going to be my cage-fight.

One of the protestors was trying to break the large plate glass window of the Whole Foods store on Telegraph Avenue. I tackled him to the ground and got him in a headlock. I figured I was going to conduct a citizen's arrest of the vandal. I was quickly surrounded by about five of his fellow Antifa thugs, taunting me, but I wasn't going to let go.

At that point, one guy in plain clothes raced up to me, the Antifa thugs letting him easily pass, and said, "What are you doing? These people are going to hurt you. You need to get out of here."

I looked at him and thought to myself, *Who is this guy?* The Antifa guys were letting him intervene, almost as if he was directing the event. *How does he know people are going to try and hurt me?*

"This guy's getting arrested," I shouted.

The plain clothes Antifa leader seemed to shrug, looked at the black clad Antifa thugs, and they went after me, hitting me with their thick wooden flagpoles. I was wearing Kevlar and a motorcycle helmet. I could take a beating and still look like Superman.

A crowd had gathered to watch the disturbance. My willingness to get beaten seemed to have overcome the "bystander effect" where people are simply too stunned by an outbreak of violence to act. Many of the genuine Occupy Wall Street protestors came to my aid, pushing the Antifa thugs aside. The person I had in a headlock escaped.

But the other protestors and I formed a human chain protecting the Whole Foods store. We won the day. From what I was able to record, in addition to what others recorded on their phones, I was eventually able to stitch together a video account of the event. The video went viral and suddenly Occupy was ejected from the parks after that.

My own experience at the Whole Foods Market on Telegraph Avenue, combined with questions raised by others, made me wonder what was really going on. When Occupy rejected the Antifa attempt to infiltrate their ranks, Occupy was ejected from the parks after that.

CHAPTER FOUR

The Hand of Fate Sometimes Points in the Wrong Direction

One of the great benefits of working at Google was that after five years they encouraged you to take a month-long "sabbatical," with the idea you should explore other areas. The hope was you'd bring back some innovative ideas to the company. I started brainstorming ideas on how embedded computer electronics, like LED lights, could be used to make bike riding safer. I figured that would benefit both motorcycle riders and those concerned about the environment since motorcycles used less gasoline than automobiles. During my Occupy New York days I'd met Waylon, a costume designer, and we came up with some initial designs. I figured I'd take my month-long sabbatical and fly Waylon out to San Francisco so we could work on the project.

Waylon came to San Francisco and we developed several good designs for leather biking gloves with a turn signal on each hand. When the month was up, I had a working prototype and was considering how I might balance my job at Google with my side-gig trying to help the planet.

* * *

During that time, my friend Dustin found out Antifa was planning a march in San Francisco on Valencia Street. He decided to go and videotape it. Antifa was advertising the event in well-known anarchist publications in the Bay Area. They provided the date, the meeting place at Dolores Park, and advertised it as "A Night of Mayhem and Ruckus." People I knew noticed the ads and asked the obvious question, "What's going to happen?" as well as "I guess the police are going to be ready for it."

The Antifa mayhem and ruckus was confined to Valencia Street, from 13th to 26th. When my friend heard it was happening, he headed out with his iPhone. He tried to interview several Antifa members to get them to explain what they were doing, but few were willing to talk. As my friend passed 17th Street, he looked down the street at the police station on Valencia. All the police cars were gone, and the station was empty. There are always cars and personnel at that 17th and Valencia station. Yet there were none to be found at that time. Car windows were smashed as well as windows of stores. It seemed as if they had done a targeted hit on glass windows wherever they could find them. It's chilling that so few police were present for an announced Antifa/Anarchist meet-up which was advertised as a "Night of Mayhem" a few days in advance.

After Antifa had smashed windows and caused damage from 26th Street all the way down to 13th Street, the police suddenly appeared. What would happen? Was it the inevitable confrontation between Antifa and the San Francisco cops? Perhaps the police had executed a brilliant strategic retreat and were now going to swoop in and arrest the rioters who'd already inflicted significant property damage on the citizens of San Francisco.

But nothing like that happened. The police stood their ground at 13th Street and Antifa melted away. One could make the argument that city officials simply wanted to contain the damage. And yet, it's equally valid to suspect an amount of damage was due to something more.

* * *

After my one month Google sabbatical, working as hard as I could on my LED biking glove with Waylon, I decided I liked trying to be an entrepreneur rather than a software engineer.

At first, I figured I'd continue working at Google and focus on the glove after hours, but another wrinkle had entered my life. I'd also started a relationship with a wonderful woman and that takes a significant amount of time, especially if you want a good relationship.

I decided to talk to Google and see if I could work part-time, cutting my work schedule to three days so I could focus the remaining time on the glove, as well as my girlfriend. I thought it was a reasonable suggestion, but the woman I met with in the human resources department said absolutely not. She informed me I had to leave Google immediately. Technically, if I was working on a new product while working for Google, Google owned it.

Because of those concerns, in August 2013 I resigned from Google, after five and a half years of working for them (and I would later rejoin Google in 2016). The experience was so liberating. I'm not sure I've ever been so happy in my entire life as I was during those first weeks and months after I left Google. I woke up with such energy to work on my new product, my love life was great, and since my girlfriend was working during the day, I often took care of her dog, a cute little West Highland Terrier named Princi.

Everybody who tried my cycling turn glove thought it was one of the best things they'd ever seen, and I had little doubt I'd eventually be showing it off to Mark Cuban, Kevin O'Leary, and the other hosts of *Shark Tank.*

The internet had exploded with options for budding entrepreneurs to raise money, and I chose Kickstarter to get my needed development funds. To make my pitch as strong as possible, I hired a professional BMX cyclist to demonstrate the product. It was a humorous video, showing the cyclist doing all these dangerous tricks on the streets of San Francisco at night, from riding on the back of his wheel, the front of his wheel, and steering while popping a wheelie. But when he went to make a turn, he demonstrated how to do it safely with my turn-signal glove. His narration implied in the short video about how the real danger to a cyclist came not when he did all of the super technical wheelies, but when making a turn in traffic.

A dedicated minority of genuine cyclists pointed out a significant problem with the video. In the last three seconds, the cyclist goes to make a left turn, but because the arrow pointed the way that it did, a left turn with the product used the same gesture as a right-handed turn. The mistake I'd made was having the LED arrow go all the way down the pinky finger. By simply changing the arrow so it pointed to the knuckle of the pinky, we solved the problem.

To do this, I brought the professional BMX rider back and we reshot just that last segment. We received about seventy-four thousand dollars in pre-orders. My prototype was solid, and I simply needed to move into production. But because of a series of disasters in manufacturing, I'd end up forking over about sixty dollars per glove and selling the glove for sixty-six. I originally wanted to find an American manufacturer, but quickly discovered the entire textile manufacturing base in the United States had been completely gutted by our "liberal" free trade policy with China, which gave the country a massive tax advantage.

I needed to have the glove made in China. I got in contact with an American ex-pat in China who knew his way around the Chinese factories. He found a good candidate and I gave them a sample of what I wanted.

Holy shit!

On the first take they produced just about exactly what I wanted, and in some ways, it was superior to the sample I'd submitted. I made a few modifications, but with that problem solved I could move onto the electronics. I was pleased to conclude it made sense to manufacture the electronics portion of the glove in the United States. However, that turned out to be the biggest mistake of the entire project.

We had a manufacturing defect in that the glove would burn too much energy when the two activation plates touched.

We tried to fix the problem by altering one of the components on the computer board to make the sensor use less power, but that just made it trigger happy. If you lived in a humid climate like Florida, or tended to sweat a lot, the turn signal just turned on, and stayed on, even when the contact plates weren't touching. It was a major product defect and luckily, we hadn't sent out many gloves when we discovered the problem. Still, we had 2,600 pairs of defective gloves and had to send out a recall for those gloves we'd already shipped. Since each pair had a glove for the right and left hand with a defective sensor, that meant we had to fix 5,200 gloves. The fix required taking the glove, pulling back the fabric, keeping the material steady with the use of wooden clothing pins, removing the circuit board, and then, with a heated soldering iron, removing the resistors. Each glove required about five minutes to repair.

I got so freaked out that my entire family from Oregon came down to help with the repairs, working in shifts to get the gloves out to my Kickstarter supporters. It took about four months to get all the repairs done and shipped out. But I figured we'd made all the mistakes we were going to make and had a truly remarkable product to sell.

I placed an order for the next shipment of gloves and things started to go wrong immediately. The gloves we got had their wires cut, and the first factory sent them to a second factory to get repaired. But the second factory had trouble with the repairs, and the two Chinese factories were complaining to me in dueling emails and phone calls. Despite being warned by the factory not to go, I booked a flight to China so I could be physically present there. I spent the Christmas break of 2014 working twelve hours a day at the factory, collapsing at night at a hotel nearby, then getting up the next day and doing it all over again. But I got the problem fixed and in early 2015 we were shipping out gloves.

However, we continued to be plagued by quality control problems and various issues popping up unexpectedly. I thought I'd be making a good amount of money by that time, but all the fixes and problems were eating into my profits. And it wasn't like I was living the high life.

The labor costs were killing me, especially since I'd hired my girlfriend to do several of the jobs. Maybe some couples can take the strain of working together as well, but we couldn't. Eventually in the summer of 2015 the relationship soured, and we broke up.

While the loss of the relationship came at a great personal cost, the effect on the financial stability of my company was dramatic. I'd been paying her so much for her hours of labor that it was sinking the company. I was able to keep the company afloat for another year by doing essentially all the labor myself. I was both sad and angry about the failure of the relationship.

And not soon after my break-up came an amazing opportunity. A new reality television show put together by none other than producer Mark Burnett, the creator of *Shark Tank*, *Survivor*, and *The Apprentice*. The show was to be called *America's Greatest Makers* and would feature inventors building a product to be sold to the public.

The grand prize was a million dollars.

I might have failed in love, but I was determined to become an entrepreneur and Reality TV star.

* * *

Maybe the biggest surprise to me shouldn't have been that reality television isn't real. Yes, I figured there'd be an element of show business and drama to the effort, and I played that to the hilt. But I hadn't expected the show to be nearly as fake as big time wrestling.

Prior to doing the show, I had created an LED face mask that I called Space Face. It had 538 multi-colored PEDS that I video-mapped to music videos. I'd run these incredible VJ loops right on my face at sixty frames per second. I made it to be the most incredible "thing" anyone could have seen at Burning Man. It was now the perfect prop to get me on a reality TV show.

I brought my face mask which had 538 LEDs on it which gave off a blinding light. And I programmed it so a movie was running on my face. The mask had been a huge hit at Burning Man when I showed up with it. People just couldn't take their eyes off me when I placed it on my face and walked around. The mask made me a popular man. If I wore it at one of the many concerts at Burning Man, the band members were always inviting me on stage to dance.

In addition, I brought a prototype joining elegant LEDs with crystals that I called Lumistones. In my audition for *America's Greatest Makers*, I also brought samples of my motorcycle glove to show my diverse range. I made a critical decision that since the contestants were likely to be mostly male, it would be good to have a female partner for the "diversity and inclusion" factor. I chose a woman I knew in San Francisco, a fire-wrap artist (dancing with fire batons) and creative thinker.

When we arrived for the audition and I could see the competition I felt pretty good.

Some of these contestants were still in high school. And while some of the contestants were brilliant, that didn't necessarily translate into charisma of the kind for which the producers were looking. I felt I had a combination of being a good technical thinker and problem-solver, as well as having something of a dramatic flair. I would make good TV.

But a problem hit, right before I was to go on stage in front of Mark Burnett himself and make my pitch. The backstage staff made the surprising move to forbid my LED face mask from being brought into the audition room.

I said, "Is there any way we can let it in? It's a major prop!"

But she was firm. I went into the audition room and did my schtick about "the future of LEDs fused with crystals." I finished by saying, "And Mark, I COULD have demonstrated my technical skills with an amazing LED face mask, but sadly, it was intercepted by the backstage staff."

Mark responded: "Really? Well, in that case, let's bring it out so you can show me." At that moment, one of the staff members took off running out of the room to retrieve my LED facemask.

I felt relieved when it was put in my hands, put it on, and I flicked the switch. Looking out from the eyeholes, I could see the room light up with vivid shades or green, red, and blue light.

Mark Burnett squinted at the bright light. I started talking, but Mark immediately interrupted. "So, why not put a phone screen behind the crystal? Why do you want to put LEDs there?"

And I responded, "It's because LEDs are more efficient than the cell phone screens. Free LEDs have a typical efficiency between 21 percent to 43 percent, compared to the phone screen which has an efficiency of around 7 percent. That allows us to have a longer battery life."

Mark looked over at his technical consultant, a high-level Intel employee, and I held my breath. The Intel employee nodded and leaned over to Mark and said in a voice loud enough for me to hear, "He's right. LEDs are much more efficient."

The exchange with Mark Burnett continued and as I walked off the stage I felt my chances of being on America's Greatest Makers had gone up dramatically.

I was going to be on America's television screens for a season, showing off my creative talents, and walking off with a million dollars to fund my company.

* * *

The problems started shortly after we were picked for the show and they gave us the rules for the competition.

They told us the original product we were supposed to develop on the show needed to be built with a specific Intel chip. I thought, *Oh, Intel is really good at building chips. That's what they do. I'm sure they're going to build a chip that's really well-engineered.* Nothing could have been further from the truth.

The product they wanted our projects to be built on was the worst designed piece of shit I've ever seen. Instead of building something from the ground up to be energy efficient and powerful, all they did was take a chip from a desktop computer and shrink it down. The chip was not ready for production. It had a whole bunch of problems with the power management of the devices. I could not even hook it directly to a lithium battery like a normal device since the processor needed a minimal voltage that far exceeded the voltage of the battery.

You couldn't use a lithium battery to charge it because it needed 12.6 volts, and that was beyond the voltage of the typical lithium battery. When the chip was fully charged, it would put out 4.2 volts, but that quickly dropped to 3.0 volts. The Lumistones needed 4.6 volts to function properly. In order to get the power to the needed level, I had to wrap the chip with a magnetic coil.

The producers, seeing how much contention there was among the contestants, gave everyone ten thousand dollars to develop our own chip. The typical chip on the open market which I'd use to build our product is about fifty cents. However, the Intel chip we were required to buy cost between ten and twenty-five dollars. In addition, when you put a traditional microprocessor chip onto a circuit board it's relatively simple because there are small pins that poke out and they're easy to weld together. The Intel chip had pads and one needed to use an extremely sophisticated board fabrication technology to marry it to the circuit board. And it only worked about fifty percent of the time and cost hundreds of dollars per board.

Using the board fabrication technology each time cost two hundred dollars. But since it failed half the time, that meant the real price for the simple procedure added four hundred dollars to the development cost. I couldn't escape the feeling that not only was the show little more than a long, sustained infomercial for Intel, but they also wanted cheap, smart labor to work for hundreds of hours on the chip to flush out all the problems.

It was clear to me I couldn't develop my product with the Intel chip, because of both performance and financial concerns.

I was going to have to depart from the rules.

I used another chip to develop my product.

But the producers must have gotten wind of my rebellion, or perhaps it was something that was happening in several groups because a directive came down that there was going to be a surprise inspection of all of our products to make sure we were all using the Intel chip.

Although it was supposed to be a "surprise" inspection, they told us when it was scheduled, and mine was supposed to take place the next day. I arranged to be busy filming a section of the show on my scheduled time, so I was able to push the inspection back another day.

With the inspection scheduled two days away, I placed a call to my technician/co-founder in San Francisco and said, "Dude, we've got a big problem. They have to see the Intel chip in the product."

He drove down a day later. In the evening, we sat down with the product to swap out the chip and make sure it had enough voltage to work. The next day when they did their inspection, we pulled back the fabric, they saw their flawed Frankenstein chip, and said, "Great! Awesome!" and marked that we'd complied and were on their way.

I was able to walk onto the stage for the first challenge wearing my gloves, my mask of light, this black leather jacket with these super bright LED hundred-watt lights, clutching one of my Lumistones (that I had started calling "fashion stones") with an actual Intel chip.

I was a being of light as I walked onto the stage. I reached up to take off my mask and said, "My name is Zach and I'm from the future." I offered them my jewelry. "I've come to give you Zackees fashion stones." The jewelry looked fantastic. I also demonstrated it had an alarm, so if you were out partying late one night and found yourself in a dangerous situation you could call for help.

The judges loved it and we easily made it past the first-round elimination.

I had a sense that the producers liked me and my partner, but it wasn't until I saw the first episode that I understood how much they liked us. Approximately a third of that first episode was dedicated to us and some other teams got less than a minute. I'd received some advice from product developers that I shouldn't expect television to drive sales. But I hoped they were wrong. It turns out they weren't.

I felt we were strong in the early part of the show and couldn't see how we wouldn't end up winning that million dollars. Or at least be among the three finalists, each of whom would receive a hundred thousand dollars. A million dollars would set me up for at least five to seven years, meaning I could do everything I needed to do. A hundred thousand would set me up for at least eighteen months.

As I continued working on the show, I was convinced leaving Google was the best thing I'd ever done.

* * *

It's often said you sometimes miss the biggest things in your life as they're happening.

However, it's difficult to overlook seven foot one NBA legend Shaquille O'Neal when he shows up on the set as a guest judge. Still, I didn't realize what an enormous impact he'd have on my life.

But before I met up with Shaq, the show had us do surveys and customer interviews to determine what they thought of our product. The light-up fashion stones were very popular, especially when I detailed some of the features and improvements I'd made, but the consistent comment which came back from black customers is that they wanted it bigger.

Shaquille O'Neal has been one of my long-time idols, so if I was going to be making something for him, I wanted it to be a Zackees fashion stone fit for a king.

I know people say you shouldn't meet your heroes because they'll inevitably disappoint you. Maybe that's true, but there's no denying Shaq was having a bad day on the set. He had a heavy chest cold, which made it difficult to understand him when you were interacting with him. All the other contestants who weren't interacting directly with the panel could listen to the audio directly in their earpieces. But when we were on camera, they didn't want us wearing our earpieces.

It came time for me to make my presentation, so I pulled out the enormous Zackees fashion stone I'd made for Shaq and placed it around his neck. The necklace was too long, so it hung down to an inappropriate place. But Shaq took it in good humor saying, "I haven't worn anything like this since my Run-DMC days," referencing the urban rap group.

"According to our customer surveys, black folk like large pendants."

Other members of the panel were talking, and I knew Shaq was saying some things. But I couldn't understand because of his heavy cold. I just smiled and nodded like an idiot, as if I understood all of it, then walked off-set.

I was backstage, getting undressed, when some of the other contestants came up to me. "Man, that was brutal," one said.

"Yeah, that was really bad," said another.

I was about to ask what the problem was when members of the production crew showed up. "We need to shoot again," they told me.

Apparently, Shaq had been calling me a racist because I'd used the term, "black folks." I was trying to be friendly, familiar, and using the same words used by the blacks I interviewed to describe themselves. I meant no disrespect. I thought I was being culturally sensitive.

At least they didn't show that footage to America and we reshot the interaction. But after that, I was a dead man walking.

Even though we'd been dominating the show up until that time, we were cut in the next round, finishing just out of the money.

One of the supposed "rules" for the competition was we couldn't come with an existing product, and instead had to develop something on the show. But the grand prize winner of that million dollars won with a tooth brushing game system for kids that had already been on the market for a year.

Still, I'd been on the show for several episodes, they'd highlighted me, and I figured I could still get an enormous amount of exposure for my business. Yes, they'd given us crappy technology. Yes, I'd had one of my heroes call me a racist. Yes, they'd broken their own rules by letting a product that had already been out on the market win the competition. But I wanted to support the show.

Maybe they didn't think the finished show was very good because the producers didn't put much money behind advertising the show. It appeared on a Tuesday night on TBS and barely caused a ripple. Even though I'd been featured in that first episode, on the day after it aired I had only a hundred people visit my website (Lumistones.com) which had been prominently featured on the show. Out of those hundred visitors, only one made a purchase.

I thought the reality TV show would be a big success. But in truth, it was just a distraction. I should have listened to the people in the venture capitalist space who told me television always underperforms. And they also told me all the people on *Shark Tank* are liars. Of course, I didn't believe them when they told me. I figured they were just jealous because they'd never been on *Shark Tank* or scored it big in reality television.

I learned an important lesson. Too much optimism can be fatal. I should have been more realistic and listened to the advice of people who knew more than I did.

As I realized that no matter how much I was featured in the remaining episodes of *America's Favorite Makers*, it wasn't going to rescue my business, and I was essentially broke. I started having a nightmare that I didn't even have enough money to pay rent. I envisioned myself showing up at the apartments of friends, asking if I can stay for "just a few days."

* * *

As I was in the midst of my career worries, my sweet dear grandmother, who always loved me so, died.

I will always remember the trips to Idaho to visit with her and my grandfather and how they'd leave candles in the snow when we visited them in the winter. Her death came at the end of a long and full life. Like so many of her generation, there was something proud and self-reliant in the way she carried herself. Her eyes had seen the total transformation of the country in a way I could never fully understand, but could at least partially appreciate.

As a result of the liquidation of my grandmother's estate, I was given a gift of ten thousand dollars. That allowed me to stay afloat for about three more months, during which time I could figure out what to do with my life.

I applied to Google, Apple, and one other company. I wanted to go back to a life where I'd wake up every morning, the work would be waiting for me, and I'd make lots of money. You know, the easy street of Big Tech, rather than the mean streets of American capitalism. I had debt payments because I'd ordered another supply of the Zackees turn signal gloves, and I figured that if I went back to work, I could dig myself out of my financial hole in about three years, as well as plan the next phase of my life. During that time, I could also streamline my business (not telling my employer about my side-gig) and resurrect my dream of one day becoming an entrepreneur. I had not given up my dream of being the owner of a successful business.

Google was interested in me again.

Although there was interest from other companies, I felt most comfortable returning to Google. I was officially hired in August 2016. The position they had for me would be at the YouTube offices in San Bruno, much closer to my apartment in San Francisco than the Google corporate headquarters in Mountain View. The commute would be so much easier.

Life was shaping up to be excellent again.

I wasn't paying much attention to the presidential election as September and October rolled around. I noticed how my co-workers were all talking about how if Trump was elected, he was going to be a fascist Nazi dictator. *Yeah, yeah*, I thought. Just more mainstream media fearmongering. Honestly, I didn't pay much attention to the election. I figured Trump was basically just a carnival show.

Then November came around and Trump got elected.

And everything went nuts.

CHAPTER FIVE

Building the "Ministry of Truth"

By its very nature, free speech in a democratic society is supposed to be messy.

The idea is we'll bump up against each other and each side will present their version of the facts, as well as the importance of the information. Then we, as the public, get to determine who made the most compelling argument. Many times, I've been of one opinion, only to find I have a different view after listening to the other side of the argument. Nobody had ever "curated" the information I'd received. Each side independently decided what information to present and then I, as a citizen, concluded which side made the most sense.

It was sounding to me as if Google was going to make the decision of what information I'd be allowed, or directed, to consider.

Eric Schmidt had built Google as an "open" company, which meant any employee could access most company documents to see what other people were doing through a system called MOMA. I typed in "fake news" to see what would come up in a MOMA search.

The first thing I found was a design document talking about the big news problem in the recent election. It began by rattling off five examples of "fake news," and the amount of Facebook engagement with each story, compiled by Statista and published in *Business Insider*.[1]

The five top "Fake News" stories listed in order were:

1. "Pope Francis Shocks World, Endorses Donald Trump for President, Releases Statement." (960,000 Facebook shares)

2. "Wikileaks CONFIRMS Hillary Sold Weapons to ISIS . . . Then Drops Another Bombshell! Breaking News" (789,000 Facebook shares)
3. "IT'S OVER: Hillary's ISIS Email Just Leaked & It's Worse Than Anyone Could Have Imagined" (754,000 Facebook shares)
4. "Just Read the Law: Hillary is Disqualified From Holding Any Federal Office" (701,000 Facebook shares)
5. "FBI Agent Suspected in Hillary Email Leaks Found Dead in Apparent Murder-Suicide" (567,000 Facebook shares)[2]

One has to notice that there were four negative stories about Hillary Clinton, and one story about a papal endorsement of Trump that nobody could possibly believe (even though it might be amusing to consider for a moment).

There was also an interesting chart at the bottom of the page showing that Facebook engagement for their definition of "Fake News" from August to Election Day was 8.7 million, while the mainstream news had 7.3 million engagements. Put in its starkest terms, the data was showing Facebook users were more likely to seek out negative stories about Hillary Clinton and more likely to seek out positive stories about Donald Trump. If that was the data, why was it a surprise to the executives at Google when Hillary lost?

There was one thread I wanted to chase down, namely whether Hillary Clinton, as secretary of state under Obama, had been running weapons through a CIA safe house in Benghazi, Libya to Turkey, where it made its way into the hands of ISIS fighters in Syria. I recalled Senator Rand Paul had grilled Clinton during a senate inquiry into the September 11, 2012 Benghazi disaster when four of our diplomats, including Ambassador Christopher Stevens, had been killed. This was some of their exchange:

> SEN. RAND PAUL: My question is, is the U.S. involved with any procuring of weapons, transfer of weapons, buying, selling, anyhow transferring weapons to Turkey out of Libya?
> SECRETARY HILLARY CLINTON: To Turkey? I will have to take that question for the record no one has ever asked me.
> RAND PAUL: It has been in news reports that ships have been leaving from Libya and they might have weapons. What I would like to know, is the annex that was close by [in Benghazi]. Were they involved with procuring, buying, selling, obtaining weapons, and were any of these weapons being transferred to other countries—to any countries, Turkey included?

> HILLARY CLINTON: Senator, you'll have to direct that question to the agency that ran the annex. I will see what information is available. I do not know. I do not have information on that.[3]

The reports were initially that a "peaceful protest" in response to a YouTube video critical of Islam had turned violent. Media accounts, usually led by Fox News and the website Breitbart, noted that "peaceful protestors" generally aren't armed with grenades, mortars, and AK-47 rifles.

I decided to go down the rabbit hole to figure out if Hillary Clinton had been running weapons through Benghazi to ISIS terrorists in Syria.

According to one account, dubbed "fake news" by the Google censors, before the attack on the Benghazi compound, a US military helicopter flying in Libya had been the victim of a Stinger missile attack.[4] However, the terrorists, in their haste to bring down the helicopter, had forgotten to arm the missile. The Stinger flew straight and true to its target, putting a dent in the side of the helicopter, before falling to the desert floor. A team was sent out to retrieve the downed missile and was successful. When they examined the Stinger's identification marks, it matched a missile the military had provided to the CIA. The inevitable question the military asked was "Why is a CIA Stinger missile ending up in the hands of terrorists?"

There were a few possible answers, but there had been published accounts of what had gone on, although they were unconfirmed. Here's an account from PJ Media in May 2013:

> Stevens' mission in Benghazi, they will say, was to buy back Stinger missiles from al-Qaeda groups issued to them by the State Department, not by the CIA. Such a mission would usually be a CIA effort, but the intelligence agency had opposed the idea because of the high risk involved in arming "insurgents" with powerful weapons that endanger civilian aircraft.
>
> Hillary Clinton still wanted to proceed because, in part, as one of the diplomats said, she wanted "to overthrow Gaddafi on the cheap."

This left Stevens in the position of having to clean up the scandalous enterprise when it became clear that the "insurgents" actually were al-Qaeda—indeed, in the view of one of the diplomats, the same group that attacked the consulate and ended up killing Stevens.[5]

It could have been simple incompetence. Where had the Stinger missile been stored? Maybe they were stolen. Perhaps the CIA had given the Stinger to a friendly Middle Eastern government and somebody in the security services in that country had provided it to terrorists. We had given many Stinger missiles to Afghan rebels in their fight against the Soviet Union in the 1980s. There are plenty of non-evil possibilities for the event. In fact, even the PJ Media article took the position that it was stupidity, rather than an evil design:

> The former diplomat who spoke with PJ Media regarded the whole enterprise as totally amateurish and likened it to the Mike Nichols film *Charlie Wilson's War* about a clueless congressman who supplies Stingers to the Afghan guerrillas. "It's as if Hillary and the others just watched that movie and said 'Hey, let's do that!'" the diplomat said.[6]

And yet, one cannot avoid the implications for how such bugling could lead to unprecedented military engagements, resulting in the deaths of millions.

What would be the result of a Stinger missile attack on a military helicopter operating in Libya? Would it draw the United States into a wider war in the Middle East? Was it a planned operation by the State Department, or simply a mistake?

It made me think of a section from Plato's *Republic*, in which he discussed the remarkable power of those who tell stories. Human beings understand the world through storytelling, a narrative which holds together with its own internal logic, not facts. A powerful narrative moves people more than data. Some may think this a flaw in human beings, while others consider it a strength. Is it better to understand the broad scope of an issue, or is it better to understand each one of its individual components? This is part of what Plato had to say about the power of storytellers and how the rulers might respond to these uncomfortable truths:

> Because I think we'll say that what poets and prose-writers tell us about the most important matters concerning human beings is bad. They say that many unjust people are happy and many just ones wretched, that injustice is profitable if it escapes detection, and that justice is another's good, but one's own loss. I think we'll prohibit those stories and order the poets to compose the opposite kind of poetry and tell the opposite kind of tales. Don't you think so?[7]

Google was publicly saying it simply wanted to curate information to make sure people had the correct "facts." But stories are made up of individual "facts" and if you're choosing which "facts" are allowed in the discussion, you're changing the narrative itself. Was Google preventing information about possible "wrongdoing" by the CIA, and thus allowing it to go unexamined?

Many have wondered if Plato's *Republic* was meant ironically, or whether he was genuinely advocating what to many today would seem to be the setting up of a dictatorial state. Regardless, it seemed Google was actively trying to create a system set up according to Plato's ideas. Modern technology has given the tyrants of today a power undreamed of by the ancients. No longer would rulers need to have to deal with uncomfortable narratives. Those stories would simply not exist, or be exceedingly difficult to find. People would believe they were seeking out independent information, only to be fed the stories approved by Google.

In the Google meeting I'd watched after Trump's election, they spoke of their fear of "nationalist" movements. We all knew when they used the word "nationalist," it was code for the rise of Adolf Hitler in Germany. Trump was Hitler in their minds, leading the United States down a similarly violent road. But the analogy was misleading.

After taking power in Germany, one of the first things Hitler did was shut down dissenting voices. Hitler didn't have to worry about "fake news" stories questioning whether he had Jewish ancestry, was a closeted homosexual, had an affair with his niece, Geli Rabul, who died under mysterious circumstances, or claims he had only a single testicle. In that instance, Google was the one following Hitler's playbook, shutting down dissenting voices under the rubric of "fake news."

I looked at the story about the helicopter being hit by a Stinger missile and its recovery, as well as the inevitable questions asked, and it seemed like a solid story.

Now, could I conclude the story was true?

No, I could not.

But it had me intrigued. It was a thread I wanted to pull because it seemed vitally important. The CIA is specifically tasked with providing disinformation to foreign countries and manipulating their systems. Had the CIA attempted to use that same skill set on Americans?

I wondered why they were choosing to label the story as fake. What if it wasn't fake, just politically inconvenient to Hillary Clinton? Maybe the claim about Google being on the side of Clinton and the globalists wasn't so far-fetched.

But something was nagging me. As an engineer I understood there couldn't simply be humans making the decisions as to what was fake and real news. They had to have some sort of "fake news" filter to censor what they didn't want the public to see.

There had to be a program to accomplish their plans.

* * *

In early 2017, I found the name of the system they'd created to take down the newly elected president of the United States.

If there was an effort to describe what fake news was, then I knew the engineers at Google would have a system designed to fix it.

When I went digging around just a little bit, the censorship project leaped out at me almost instantly. I knew from the moment I saw the project name that it was a perfect censoring tool, and if you criticized it, it would be easy to categorize you as a bad person.

The name of the program was **MACHINE LEARNING FAIRNESS**.

Machine learning is a type of artificial intelligence which uses simulated human neurons to try to find patterns in piles of data. Once it gets trained it can sift through data and make decisions on what classification to give the information. For example, let's say there's a certain way opponents of Hillary Clinton often refer to her, maybe something like "Crooked Hillary." Anybody who uses that expression might have their post or article immediately flagged as a possible example of "fake news." The decision can then be made whether to rank that post high in results or low.

Or let's say you teach the system to flag any combination of the words "Trump" and "Stormy Daniels," then you can make a similar decision on how to rank the post in search results. It's an amazing power but carries with it the risk that conversations do not happen organically. They are manipulated. However, no search engine could advertise itself by saying, "Use our search engine and we'll teach you the right way to think!" Even the most liberal thinker wants to believe their opinions are the result of their own independent judgment and discernment of information.

I was amazed by the breadth of the documents. There were artificial intelligence systems for YouTube that were decoding all the human audio into words to figure out whether somebody was saying a bad word or using profanity. But the system could also be configured to determine the political leanings of the speaker by their choice of words.

There was a power point presentation on Machine Learning Fairness and algorithmic unfairness, and why we needed to solve it. One of the points was that humans are fundamentally racist, and we need to correct for that or else our data is going to reflect our internal biases and those biases are going to amplify injustice. Many of the claims seemed to me to have a culturally Marxist flavor to them.

For example:

> **If a representation is factually accurate, can it still be algorithmic unfairness?**
>
> Yes. For example, imagine that a Google image query for "CEOs" shows predominantly men. Even if it were a factually accurate representation of the world, it would be algorithmic unfairness because it would reinforce a stereotype about the role of women in leadership positions. However, factual accuracy may affect product policy's position on whether or how it should be addressed. In some cases, it may be appropriate to take no action if the system accurately reflects current reality, while in other cases, it may be desirable to consider how we might help society reach a more fair and equitable state . . .[8]

In my mind, I was torn. On one hand, I was thinking, "holy shit, this is a radical policy. They're deciding what part of objective reality they want to change." What if they decided it was prejudiced to show only tall basketball players?

On the other, I was saying, "Well, they're a private company. They can be as messed up as they want to be." I continued to read on, finding the definition of algorithmic unfairness as "unjust or prejudicial treatment of people that is related to sensitive characteristics such as race, income, sexual orientation or gender, through algorithmic systems or algorithmically aided decision-making."[9]

Despite the fact they were a private company, they were going to be enforcing this on the rest of the country and that was a big red flag in my book. I knew about anti-trust violations. In short, you need to be BIG, you needed to be BAD, and you needed to be BOTH.

Google was openly declaring war on objective reality, according to their own internal documents. And without disclosure, they were going to roll this monstrosity out to the American public. The document set out their goals with regards to algorithmic unfairness:

Our goal is to create a company-wide definition of algorithmic unfairness that:

1. **Articulates the full range of algorithmic unfairness that can occur in products.** This definition should be robust across products and organizational functions.

2. **Establishes a shared understanding of algorithmic unfairness** for use in the development of measurement tools, product policy, incident response, and other internal functions.
3. **Is broadly consistent with external usage of the concept.** While it is not a goal at this time to release this definition externally, it should represent external concerns so that we can ensure our internal functions address these concerns.[10]

I know I'm a software engineer and sometimes I don't communicate things clearly, but that seemed pretty simple and terrifying.

First, they wanted to create a definition of "algorithmic unfairness."

Second, they wanted to put that definition into practice throughout Google.

And third, they quite clearly didn't want to tell the public what they were doing.

Further on in the document they listed several areas of concern. One of them was "Revamping News Corpus," or in plain English, the "body of news."

> **Goal:** Establish "single point of truth" for definition of "news" across Google products. Mitigate risk of low-quality sources and misinformation in Google News corpus.
>
> **Status:** Define new Google News corpus utilizing existing infrastructure and tools while integrating different quality tiers and labels, algo-human content review, new cross-product review/exclusion pipeline and updated inclusion . . .
>
> **NEXT STEPS:** Research and gather info about possible signals to leverage for product teams to make source quality decisions. Finalize and test new rater inclusion guidelines with standards to combat misinformation.[11]

There was a lot to unpack in that section, mostly because it was all such a naked power grab for control of information. Google wanted to establish in their words a "single point of truth." What could be more terrifying?

In addition to setting up the rules of the game, they were also choosing the referees.

There would be "different quality tiers and labels," as well as "algo-human content review." That meant the information and the people who might comment would be ranked on their reliability. In other words, they were deciding which voices would carry weight, and which would not.

"Possible signals to leverage for product teams to make source quality decisions," meant they'd be using all of these data points to make decisions on the reliability of the information. And lastly, "Finalize and test new rater inclusion guidelines with standards to combat misinformation." That's like checking the calls of referees in a college sports game in order to see if they'll do what you want for them. Even the mafia never came up with a system for total control as complete as what Google had planned.

A little further in the document I came upon the "Purple Rain" subsystem (and couldn't believe how close its intention was to the song's lyrics of telling a woman to close her mind and do as she was told).

> **Project Purple Rain: Crisis Response & Escalation**
> **Goal:** Establish and streamline news escalation processes to detect and handle misinformation across products during crises. Install 24/7 team of trained analysts ready to make policy calls and take actions across news surfaces including News, News 360 and Feed.
> **Status:** SOS alerts, Crisis Response, HotEvent, and T&S Incident Management teams are collaborating to identify a narrow set of queries that would be used to manually trigger flight-to-quality in Search. T&S Incident Management team is currently looking to expand and share resources with teams that currently handle Suggest and WebAnswers escalations.[12]

Yes, Google didn't think we could handle the news. They were going to be on guard against misinformation like superheroes with their team of "trained analysts ready to make policy calls and take actions!" And with their "narrow set of queries" which altered them to misinformation they could take to the air like Batman in the Batplane to trigger their "flight-to-quality in Search." Almost sounds like the ad a pest control company might use to rid your house of ants. However, in this case, the ants were people who simply had a different take on current events.

The next section of the document was entitled "Expanding Collaboration for News Quality," and began with some questions. There was a description of the "News Ecosystem" and "Google Efforts to Address Fake News: What Google is Doing to Address the Problem."

With pictures and text, the slide deck had the following words: "Users have access to a large variety of news Sources. Many users access news sites via FB News Feed, but some bad actors came into play. And the issue gained notoriety. We discovered that some of these sites were using AdSense to monetize their traffic. With this update, we disallow sites that mislead users."[13] Google was saying that people were making bad choices, and they were going to financially punish those who offered bad choices to their customers by demonetizing the sites. In addition, they were going to take action to make certain that advertisements were of the appropriate quality.

What followed next was a slide deck presentation which was entitled "Fair is Not the Default." The examples they used were underwhelming to me. A picture of tourists at San Marco Square in Venice, Italy began with the observation that everybody's head was at about the same horizontal level, and it was because the photographer was of about average height. If he was taller or shorter, it might look slightly less uniform. Then there was the following words spread out over several pages:

> There is talk about the role of humans in machine learning. But it's really a talk about the role of humans in decision making. It's true they can follow instructions at superhuman speed, with superhuman fidelity and over unimaginable quantities of data. But these instructions don't come from nowhere. Although neural networks might be said to write their own programs, they do so towards goals set by humans for human purposes. If the data is skewed, even by accident, the computers will amplify injustice.[14]

I quickly found the flaw in their thinking. Humans are imperfect. I confess that freely and without shame. But in our conversations, listening to voices with different points of view and different data, we improve the quality of our thinking. That's the guiding principle of Western civilization, which is why we guard the right to free speech so zealously.

I don't know the truth.

You don't know the truth.

But if we talk, perhaps we can discover the truth together. They were short-circuiting the essential human dialog which has held sway among leading thinkers since the Enlightenment of the seventeenth and eighteenth centuries, whose guiding principles were happiness, reason, nature, progress, and liberty.

There was a discussion of photo-editing and how it was standardized to make photos of people look whiter. As I read, I agreed, that's wrong. Then it started to go into some serious mind control, showing what they wanted the feedback loop to look like. "Training data are collected—Algorithms are programmed—Media are filtered, ranked, aggregated, or generated—People (like us) are programmed."[15] After that series of slides, there were some others which stated, "Unconscious bias affects the way we collect and classify data, design, and write code," and then "Unconscious bias gets reinforced in the training data."[16]

There was an example of how crash test dummies were universally male until 2011, causing females to suffer greater injuries in car crashes. Another example was color calibration of skin tones from Kodak film in the 1990s. And how in superhero movies the male characters talked more than the female characters.

The final few slides of the display seemed both over the top and chilling to me:

> We can't remove human perception from the loop. And we can't be gripped by inaction, either. The inequity demonstrated in these examples may feel overwhelming, perhaps even a little disheartening. But we're in the right place at the right time and in the right industry to do something about it.[17]

I didn't have the sense of panic that the makers of the document appeared to feel. Were there problems which needed to be address? Certainly. But in each of those examples, free speech had either solved the problem, or was being brought to bear on the issue.

They hadn't made the case to me that there was something fundamentally flawed with free speech. It just seemed like they didn't approve of the election results.

* * *

I'd read several dystopian books of the future, like *1984*, *Fahrenheit 451*, *Animal Farm*, and *Brave New World*. They were all variations on a theme, that of the destruction of objective reality in the service of some ideological vision. If you destroy objective reality, then it becomes possible to proclaim war is peace and freedom is slavery.

But the unfortunate truth is that such examples aren't limited to the pages of science fiction books. I had many friends from Russia (and the father of my close friend, Andrew, who was thrown in a gulag and wrote about it after his defection), and they told me what it was like living under the old Soviet system. The government was so top-heavy and bureaucratic that it created a circus show, a clown world where nothing worked as it was expected. It wasn't simply that there was waste and corruption everywhere. The most corrupted part of that life was the narrative, the objective reality promulgated by the State, to which the citizens were supposed to pledge their allegiance. In private, most hated the system. The only people who believed in the system were the snitches who'd inform on those who didn't exhibit the proper amount of ideological zeal.

And I thought, *Oh my, God, communism is coming to the United States and it's going to be brought to us by Google.*

That sent me into a deep funk for several weeks.

Because I love the freedom in America to be a contrarian. I used to think that's what Google and Silicon Valley stood for. Create something the world has never seen before. But that ideal was being betrayed right before my eyes.

And yet it was being done by Sergey Brin, an immigrant who'd fled that very same totalitarian communism. Did he miss the old days of his youth when he could only read the party newspaper or watch State television?

Why did he want to create a system which was the exact mirror of what he'd fled?

CHAPTER SIX

The Covfefe Deception

On May 31, 2017, after returning to the White House from visiting Saudi Arabia, the Palestinian Authority, Israel, Italy, the Vatican, Belgium, and Sicily, President Trump tweeted out:

> Despite the constant negative press covfefe.

As NPR recounted later in the year:

> The tweet was posted at 12:06 am ET and immediately became an internet sensation because it didn't make any sense. Some wondered whether the president was OK or whether he had just fallen asleep midtweet. The mysteries of covfefe were never solved. Then-press secretary Sean Spicer barely even tried to explain it, telling reporters, "I think the president and a small group of people knew exactly what he meant." The tweet was deleted a few hours later and yet, remarkably, the error of a tweet remains the president's third-most retweeted post of 2017.

A few hours later a follow-up tweet was sent which read:

> Who can figure out the true meaning of "covfefe"??? Enjoy![1]

Yes, the tweet was cryptic, but Google had a nifty translation application and when one put the word "Covfefe" into the application, it noted covfefe was an Arabic word which meant, "we will stand up!" Other internet researchers suggested it was an even more ancient Biblical word and the translation was actually "we will stand up to the fallen." (See pages 5 and 6 in the photo insert.)

Thus, within a few hours, internet sleuths had narrowed the meaning down to two possibilities. Either, "Despite the negative press, we will stand up!" or "Despite the negative press, we will stand up to the fallen!"

The next day the *New York Times* took direct aim at this interpretation in an article by Liam Stack:

> The internet is full of confident people who do not know what they are talking about. The latest example: a conviction spreading in right-leaning social media communities that a garbled tweet by the president—he wrote "covfefe"—was not a late-night typo but was instead Mr. Trump sending a message to the world in Arabic.
>
> "Covfefe," these people on the internet insist, is Arabic for "I will stand up." This is not even close to true.[2]

The source for the claim of the *New York Times* article was Professor Ali Adeed Alnaemi, a professor of Arabic at New York University, who had previously worked for the *New York Times*.[3]

> But what did the President mean? Mr. Alnaemi said the word "covfefe" was "something meaningless" in Arabic, a language that Mr. Trump, who campaigned on a pledge to ban Muslims from the United States, has never publicly claimed to speak.
>
> There is no standardized method for rendering Arabic words in Latin script, but the professor said if Mr. Trump had wanted to write "I will stand up" in Arabic he would have written something like "safeq" or "sawfa aqef."[4]

When the story broke I immediately went into the Google internal search system (MOMA) to see if I could figure out what was happening. I immediately found a document which addressed the controversy and downloaded it. The Google translate function identified the word as being Arabic, but Google was going to override its own system to say it was a mistake. The document was entitled "'Covfefe' Translate Easter Egg."

> **GOAL** We currently translate the query "cov fe'fe" from Arabic to English into "I will stand up." This created some confusion this week as users tried to translate Donald Trump's tweet from Wednesday night which had the word

> "covfefe" in it. Since the word has no real meaning, we want to do an Easter egg that translates "cov fe'fe" and "covfefe" into "(¯_(ツ)_/¯)" on Translate properties.[5]

In the computer and gaming world an "Easter egg" has a special meaning. It's usually a hidden feature in a commercially released product.

Google wanted to make sure the word "covfefe" vanished from memory.

I also uncovered another document which terrified me. On that same day, June 1, 2017, engineers at Google got to work on implementing the new translation of "covfefe" into a meaningless emoticon. That issue was reported internally at Google at 10:33 p.m. and "Assigned to derrida-team@google.com."[6]

Now what is the Derrida team? I'll tell you what I think it is.

Jacques Derrida is a French philosopher, best known as the father of "deconstruction" theory.[7] This is how *National Review* recently described deconstruction theory, which many consider to be the ideological underpinning of our current "woke" culture:

> It consists of critiquing the writings of past authors, especially male ones, 'deconstructing' them, which means exposing the submerged ideology of power, racism, misogyny, repression, and so on that is hidden below the overt text of a novel. This French cultural product, which began to occupy a prominent place in American university literature departments in the 1970s, has had the effect, over several student generations, of bringing literature departments, especially those of foreign languages, to extinction.
>
> Why? It is in the DNA of adolescents, even of those who have never heard of Jacques Derrida, to deconstruct, to tear apart the assumptions of their forbearers. When professors stopped talking about Milton's prose and began pointing out the treatment of his daughters, students got the point immediately.[8]

By definition, deconstruction *is* an instance of critical thinking, rejecting your own culture in terms of oppressor and oppressed. I think the Derrida team was created to assist in the destruction or modification of problematic words.

At 10:35 a.m. the initial issue was answered: "People are very creative in interpretation. I'll prepare a response for this right now."[9] At 10:48 a.m., with a full ten minutes of working on the problem, he wrote, "The auto-transliteration query almost matches the phrase mentioned in the article, usually

Latinized as "sawfa aqef." Since they're slightly different, I'm going to bad the misspelling."[10]

There were some bumps along the road to removing covfefe from human consciousness. On June 5, 2017 at 10:52 a.m., the original issue reporter emailed, "When will this be live?"[11]

A second engineer replied at 11:12 a.m., "I think this is already pushed out to nmt servers. Verified through modulez and prod debug frontend in IS. It might require a cache clear."[12] A cache can best be thought of as a software component that stores data so that future requests for that data can be served faster.

A minute later, the original reporter emailed back at 11:13 a.m., "The original report still reproduces,"[13] which meant the original translation, "we will stand up," was still being presented to the public.

There were some additional emails, with the original reporter assigning the problem to another employee, who wrote back at 11:33 a.m., "Clearing the cache. It may take up to 60 minutes."

And with that, the word "covfefe" which had existed for perhaps thousands of years, was wiped from the Google servers, existing only as a digital ghost.

* * *

Perhaps it was just a coincidence that on June 7, 2017, the mainstream media started questioning Trump's sanity and using this as the basis to remove him from the presidency via the Twenty-Fifth Amendment:

> Donald Trump's presidency has prompted early and widespread speculation of its end through resignation, removal or a finding of presidential inability. Whatever the plausibility or merits of such scenarios, each would involve the 25th Amendment to the Constitution, which makes it clear that the vice president will take over in any of those events and, following a resignation or removal, would also nominate his successor.[14]

What in the hell was going on here? This was an open discussion about how to remove a president of the United States. Had there really been "early and widespread speculation" of the end of the Trump presidency? Maybe in liberal newsrooms, but not across America. After more than two hundred years of presidential elections, people were used to the process. You had an election. It went your way or it didn't. And then four years later you got another chance.

The article went through the various sections of the amendment, the updating of the Constitution by the Amendment, the president and Congress sharing the power to pick a new vice president, and the ability of disabled presidents to temporarily cede power and duties to their vice president. But it was the last section, section four, which comprised the bulk of the *Washington Post* article. That was the section which "empowers the vice president and Cabinet to declare a president incapacitated."[15] The Twenty-Fifth Amendment gave a clear blueprint for how to execute a political coup. You just had to get the vice-president and a majority of the major cabinet officials on your side. The article detailed how such a plan would be executed:

> Under current law, the vice president and a majority of "the principal officers of the executive departments"—which the legislative history makes clear are essentially the Cabinet officers listed in the line of presidential succession—may declare the president incapacitated by a written notice to the speaker of the House and the president pro tempore of the Senate. At that point, the vice president automatically takes over presidential powers and duties as acting president.
>
> This section seems most likely to be used when there's an unexpectedly unconscious president—although it clearly applies if a president is incapacitated from some other mental or physical inability.[16]

Are you understanding the "wink-wink" function of this article? Let's think of Washington, DC as a place filled with self-important people who want to believe they know better than the public and only they can change history. What changes history more than replacing a president?

Did anybody take the bait?

They did.

How do we know? Because the *New York Times* told us they did.

This is from an opinion article published on September 5, 2018 penned by an anonymous author, supposedly a high-ranking official, with the title, "I Am Part of the Resistance Inside the Trump Administration." After going over numerous policy differences with the president, such as his approach to Russia and North Korea, and the president's penchant for asking difficult questions, the author noted that many of the members of the administration were actively undercutting his directives. The author defended these actions by stating:

> This isn't the work of the so-called deep state. It's the work of the steady state.

> Given the instability many witnessed, there were early whispers within the cabinet of invoking the 25th Amendment, which would start a complex process for removing the president. But no one wanted to precipitate a constitutional crisis. So we will do what we can to steer the administration in the right direction until—one way or another—it's over.[17]

When I read that article, it made my skin crawl. My suspicion that Google, along with others, had been preparing a coup in 2017, particularly after the "covfefe" incident, seemed to be right on the mark. No tyrant in history has ever ridden to power claiming they want to do bad things. They always present themselves as the "savior" of the nation.

Therefore, they aren't the "deep state," because that would be bad.

Instead, they are the "steady state," because we know steady is good.

Let's just split the difference and call them the "status-quo state." War and hostility in the Middle East have been the status quo for centuries. The system is set up to deal with that tension. Why was Trump mucking around with a system that had been so profitable for so many for so long?

Even in 2020, four years after the events, the answer was unclear, although there are many claims that Trump was messing with the status quo. This is from a *Washington Post* article from June 3, 2020 on the issue:

> It's one of the most contentious alleged scenes of the early part of Donald Trump's presidency: The idea that high-ranking officials in his administration at one point talked about invoking the 25th Amendment to try to remove him from office . . .
>
> On Wednesday, former deputy attorney general Rod J. Rosenstein shed some light on the whole thing. While testifying to Congress, Rosenstein was pressed on reporting and claims by deputy FBI director Andrew McCabe. The allegation is that Rosenstein in 2017 had discussed wearing a wire to record his conversations with Trump and spoke openly about whether the 25th Amendment could be used to get him out of office.[18]

Who were the members of the Cabinet who were allegedly working behind the scenes to invoke the Twenty-Fifth Amendment to remove the president?

* * *

While I was still in the fog of war in June 2017, trying to figure out what was going on with covfefe and Google's attempt to remove it from human memory, there was something said by the CEO of my company, YouTube (owned by Google), Susan Wojcicki, at an event in Los Angeles, CA. Her brief talk, just a little more than two minutes (she talks fast), made a distinct impression on me. It solidified my suspicion that Google, and all its assets, had turned to the dark side.

You may ask: who is Susan Wojcicki?

This is her entry in *Encyclopedia Britannica*, preceded by the announcement, "Meet extraordinary women who dared to bring gender equality and other issues to the forefront."[19]

> Wojcicki's father was a physics professor at Stanford University, and her mother was a teacher. She grew up in the Stanford, California, area and later studied history and literature at Harvard University (A.B., 1990), economics at the University of California, Santa Cruz (M.S., 1993), and business at the University of California, Los Angeles (M.B.A., 1998). After returning to Silicon Valley in 1998, she rented out garage space in her Menlo Park home to the newly incorporated Google, Inc., which briefly used it as the company's first headquarters office.[20]

Talk about good timing! Who wouldn't want to be the person who gave Google their first office space in your garage? Can you be any more "ground floor" than that? Wojcicki would go on to work in several capacities in Google before her sister, Anne (who founded the genetics company 23andMe), married Sergey Brin in May 2007 and bore him two children.[21] Anne and Sergey would amicably divorce in 2015, share the upbringing of their two children, and reach an undisclosed settlement of Brin's thirty-billion-dollar net worth at the time.[22]

Get the family connections?

Susan Wojcicki is Sergey Brin's ex-sister-in-law and aunt to Sergey's two children. Sergey runs Google, Anne runs 23and Me, the world's premier genetic testing company, and Susan runs YouTube. And who says the world is ruled by a few powerful families, connected by blood and marriage? For example, did you know that Nancy Pelosi's brother-in-law, Ron Pelosi, was married for fifteen years to California Governor Gavin Newsom's aunt, making the speaker of the house an aunt by marriage to the California governor?[23] Or that Governor Newsom's father, William, served as the financial advisor to San Francisco billionaire Gordon Getty and handled all responsibility for the Getty family businesses?[24] What can you grow up to be when daddy works for a billionaire and your aunt runs Congress? The governor of California at the very least, and who knows how high you can rise with movie star looks and good hair? As George Cariln stated: "It's a big CLUB and YOU AIN'T IN IT!"

But let's get back to the Brin/Wojcicki axis of power.

That summer, Susan Wojciki hosted a Youtube event called "Stream." This all-hands meeting was held in Los Angeles, CA for all YouTube employees. At the end of the event, Susan appeared on stage and spelled out how Google was going to implement a new program to boost up their "authoritative content." This is what she said:[25]

> The second area, fake news, it's a hard area. You know, a year ago, we didn't have the term 'fake news.' Now, we hear about it every single day. It's had a lot of concerns about how it affects elections, how it affects politics. News is important to us. We're a platform with global distribution. We talked about the number of users that are coming to us. We have some responsibility to make sure we're delivering the news when something important happens in the world.
>
> When there's a crisis we think that people could benefit from this news. So, news has always been important to YouTube. We also see people in locations like Syria where the traditional news organizations can't get to. And people are talking about citizen journalists coming out of Syria.
>
> So, what are we doing?
>
> Basically, this sounds easy, but it's really hard to do. We're pushing down the fake news. We're demoting it. And we're increasing the authoritative news and promoting it.
>
> How do we do that?
>
> We came up with trashy news, where we have classifiers where we identify it. We look for salacious clickbait content that isn't, that we don't think is, you know, the authoritative news. It's just kind of encouraging people to look at it, but it's not true.
>
> We're training. We've added these instructions to our readers, and we've updated our classifiers, and we're working to understand and identify with ***machine learning*** and then push that down.
>
> And then we're increasing our authoritative news. We're doing that with things like a 'breaking news' shelf. And we're testing it in the US and France and the UK and more countries coming soon. We have sources [stories?] that come from reputable sources. We work with Google News on that, to define what those respectable sources are. It triggered last week in the London Bridge attacks. It's also going to trigger on search when you type in something for a news event. You're going to see news there.
>
> We're also working with a lot more news publishers. News publishers don't want to be in the technical business of running their own player. We can do that

for them. And we also want to get more news players on the platform. We have the goal of getting over a hundred new news providers on our platform this year. [bold and italics added by author][26]

* * *

Wojcicki talked fast and when you listen to her it can sound like the smooth presentation of a CEO at the top of her game. You might even find yourself nodding along. Sounds good, right?

A couple problems come to mind, however.

First, this idea of "fake news" affecting elections is a little problematic. Who determines what's fake and what's authoritative? The answer for going on three centuries has been to let everybody speak, and that which is true will rise to the top. Google, and its subsidiary, YouTube, were now looking to put their finger on that scale. Some news won't "exist" because you can't see it.

Second, what's up with Syria being mentioned again? Conflicts are raging across the globe, but our generals like Mattis, and now YouTube, seemed to be highlighting this one in the Middle East, with us looking down the barrel of a gun at Russia, the other major global nuclear power.

Maybe a coincidence. Maybe not.

Third, they're going to "push down" fake news? What is it they say about repression? The more you push it down, the more it's going to pop back up. Might it be better to simply let "fake news" (if it's that) die a natural death?

Fourth, I couldn't help but feel the blood in my veins run cold when I heard that expression "machine learning" again. Why were machines going to tell people what to believe?

Fifth, hasn't pretty much every tyrant in history tried to create their own authoritative news sources?

Did YouTube and Google believe the tyrannical system of control they were creating would end up any better than all those which had failed throughout human history?

Was there something in the water of Silicon Valley, or maybe in the enlightened and superior intellects of California, which would allow them to avoid the mistakes of the past, and finally usher in the Golden Age?

* * *

On September 15, 2020, a signing ceremony for the "Abraham Accords" took place at the White House. This was a peace deal between Israel, the United Arab Emirates, and Bahrain, essentially ending decades of the cold and occasionally hot war between Israel and two of its Arab neighbors. As reported by the *Voice of America*:

> "These visionary leaders will sign the first two peace deals between Israel and the Arab state in more than a quarter century," Trump said. "In Israel's entire history there have previously been only two such agreements, now we have achieved two in a single month."
>
> Israeli Prime Minister Benjamin Netanyahu, UAE Foreign Affairs Minister Abdullah bin Zayed and Bahrain Foreign Minister Abdullatif bin Rashid al Zayani signed the so-called "Abraham Accords" on the South Lawn of the White House.
>
> "This peace will eventually expand to include other Arab states, and ultimately can end the Arab-Israeli conflict, once and for all," said Netanyahu.[27]

Was this the direct result of President Trump's speech in Saudi Arabia where he called upon the "three Abrahamic faiths" to "join together in cooperation, then peace?"

As Trump might have written in a tweet, it seemed that "despite the constant negative press," and the significant meddling of Google and YouTube, the leaders of the Middle East were finally "standing up" for peace.

CHAPTER SEVEN

The Las Vegas Massacre

This is what the mainstream media wants you to believe about the Las Vegas Massacre of October 1, 2017, as recounted a year later on the History Channel website section entitled, "This Day in History."

> On the night of October 1, 2017, a gunman opened fire on a crowd attending the final night of a country music festival in Las Vegas, killing 58 people and injuring more than 800. Although the shooting only lasted ten minutes, the death and injury tolls made this massacre the deadliest mass shooting in U.S. history at the time of the attack.
>
> Stephen Paddock, a 64-year-old retired man who lived in Mesquite, Nevada, targeted the crowd of concert goers on the Las Vegas strip from the 32nd floor of the Mandalay Bay hotel. He had checked into the hotel several days before the massacre.[1]

That's a pretty good start. Let's go over the facts. The massacre did take place on the date and at the location described. Besides the massacre of approximately 150 Native Americans by the US Army at Wounded Knee in 1890, it's accurate to call this the deadliest mass shooting in American history.

Stephen Paddock was a sixty-four-year-old retired man who lived in Mesquite, Nevada and his dead body was found in the hotel room after the massacre. Paddock had also checked into the hotel several days before the massacre. No arguments about the facts so far.

The article continued:

> Paddock began firing into the crowd at 10:05 p.m. using an arsenal of 23 guns, 12 of which were upgraded with bump stocks—a tool used to fire semi-automatic guns in rapid succession. Within the 10-minute period, he was able to fire more than 1,100 rounds of ammunition . . .
>
> An open-door alert sent hotel security guard Jesus Campos to investigate the 32nd floor at the start of the shooting. After arriving on the floor via the stairs, Campos couldn't get past a barricade blocking the entrance so he used the elevator instead. While walking through the hall, he heard a drilling sound coming from Paddock's room and was shot in the leg, through the door.
>
> Once authorities were alerted, they arrived at Paddock's suite at 10:17 p.m. and didn't breach for nearly another hour at 11:20 p.m. Paddock was found dead by a self-inflicted gunshot wound to the head. His motives remain unknown.[2]

So far so good. This is how the situation appeared at first to the authorities. However, that's not the end of the inquiry. It's just the start. First of all, who was Stephen Paddock? We know he was sixty-four years old, wealthy, and had no history of violence. How many sixty-four-year-olds suddenly decide to become mass murderers, especially with no previous history of violence? The number is exceedingly small.

The article concludes by saying that "Authorities concluded that Paddock had no connections with terrorist groups such as ISIS and that his planned attack was carried out without accomplices."[3]

Fun fact: All the cameras in the casino and the hotel were disabled! How is that possible????

The mystery of Stephen Paddock remained.

* * *

Subsequent stories in the media gave more detail, but still left many questions unanswered.

An official police timeline of the shooting, published by ABC News on October 4, 2017 and then updated on October 9, 2017, gives the following account.

> **9:59 pm:** Suspect shoots security guard through door.
>
> **10:05 pm:** First shots fired by the suspect into crowd.

10:12 pm: First officers arrived on the 31st floor and announced gunfire coming from above them.

10:17 pm: First two officers arrive on the 32nd floor of the hotel where Paddock's room was.

10:18 pm: Security officer tells the officers he was shot and gives the exact location of the suspect's room.

10:26–10:30 pm: Eight additional officers arrive on the 32nd floor and begin to move systematically down the hallway, clearing each room and looking for injured people.

10:55 pm: Eight officers arrive at the stairwell in the opposite ends of the hallway nearest the suspect's room.

11:20 pm: The first breach was off and officers entered the suspect's room, where he was seen lying on the ground.[4]

In an event of this nature there will always be inconsistencies. But the timeline has remained fairly stable. The only significant change to the account was for security guard, Jesus Campos. Initially, it was believed he was shot near the end of Paddock's rampage. That was amended to him being shot six minutes before Paddock started shooting into the crowd.

We have a time frame of approximately ten minutes of shooting, in which a sixty-four-year-old man fired off more than eleven hundred rounds of ammunition. Many would question whether a man of that age could physically fire off that many rounds, even with semi-automatic weapons.

A week after the massacre, the *New York Times* published a long piece on Stephen Paddock and his background. This is how it opened:

> Stephen Paddock was a contradiction: a gambler who took no chances. A man with houses everywhere who did not really live in any of them. Someone who lived the high life of casinos but drove a nondescript minivan and dressed casually, even sloppily, in flip-flops and sweatsuits. He did not use Facebook or Twitter, but spent the past 25 years staring at screens of video poker machines.
>
> Mr. Paddock, a former postal worker and tax auditor, lived an intensely private, unsocial life that exploded into public view on Sunday, when he killed 58 people at a country music festival and then shot himself. But even with nationwide scrutiny on his life, the mystery of who he was has only seemed to deepen.[5]

Is this the profile of a mass murderer? A gambler who doesn't take chances? A long-time government worker for the post office and the IRS?

> Stephen Paddock began buying and refurbishing properties in economically depressed areas around Los Angeles, teaching himself how to put in plumbing and install air conditioning. By the late 1980s, "we had cash flow," said Eric Paddock, who added that he had given his life savings to his older brother to invest and eventually became a partner in his company, because "that's the kind of guy he was. I knew he would succeed."

"He helped make my mother and I affluent enough to be retired in comfort," he said.[6]

Paddock was the kind of guy who made things happen. He wanted to be rich, taught himself about real estate, and then learned how to install plumbing and air conditioning. As a young man, he was recognized for his intelligence and humor, as recalled by Richard Alarcon, a former Los Angeles city councilman.

> Mr. Alarcon took a science class with Mr. Paddock and remembered him as smart but with a "kind of irreverence. He didn't always stay between the lines."
>
> He recalled a competition to build a bridge of balsa wood, without staples or glue. Mr. Paddock cheated, he said, using glue and extra wood.
>
> "Everybody could see that he had cheated, but he just sort of laughed it off," Mr. Alarcon said. "He had that funny quirky smile on his face like he didn't care. He wanted to have the strongest bridge and he didn't care what it took."[7]

It can be genuinely difficult to understand why people do certain things. For some actions, the reason may never be known. But when unusual patterns present themselves it's important to dig deeper.

In the final part of the *New York Times* article, they detailed how Paddock would go to certain hotels for long gambling splurges, once staying at a Las Vegas hotel for four months straight and gambling the entire time. His favorite game seemed to be video poker, which, according to the article, requires the gambler to know the history of a particular machine.

It seems unlikely that Stephen Paddock was a serial electronic machine gambler, though. Other evidence paints a theory that was easier to swallow. Stephen Paddock was a weapons dealer working with some three letter US government agencies and was using the casino as a money laundering operation for his payouts. Basically, it worked like this; first you sell weapons, then you use the money to buy chips, gamble, and then exchange the chips for money. An article from the casino trade magazine outlined this type of scheme.[8]

But the *New York Times* claimed Paddock was very good at electronic gambling and apparently, "knew the house advantage down to a tenth of a percent."[9] But none of that explains his terrible shooting rampage. It does raise a question of whether he was just that good at gambling, or whether he might have had some undisclosed sources of income that he could put at risk by such actions. The article describes him as a "mid-level high roller, capable of losing $100,000 in one session, which could extend over several days."[10] That may be true, but still raises suspicions, which have not been answered to this day. If not a violent personality, or money problems, then what motivated him?

> As for the mystery of why Mr. Paddock would go on a shooting rampage at the Mandalay Bay Resort and Casino and then kill himself, most in the gambling industry do not believe it had anything to do with money.
>
> He was in good standing with MGM Properties, the owner of the Mandalay and the Bellagio, according to a person familiar with his gambling history. He had a $100,000 credit limit, the person said, but never used the full amount.[11]

If it didn't have anything to do with money, and Mr. Paddock didn't seem to have a violent personality, what can explain this monstrous behavior, taking the lives of fifty-eight people and injuring more than eight hundred?

Yes, there were many mysteries about Stephen Paddock. But there was one fact, perhaps more important than any, that the media accounts got wrong.

Many of the most important events may not have taken place at the Mandalay Bay Hotel at all.

* * *

As detailed in an article by the *Las Vegas Review Journal*, the Mandalay Bay Hotel is more properly thought of as two hotels. The Mandalay Bay Hotel and the Four Seasons Hotel.

You may think you've fallen down the rabbit hole, but here is how this oddity was explained in the article:

> The 43-story building had an unorthodox numbering system even before the shooting. Mandalay Bay elevators showed stops at floors 1 through 34 and 60 through 63. There was no 40 through 59.

> Floors 35 through 39 are managed by the Four Seasons hotel and have a separate elevator.
>
> The Mandalay Bay elevators will now indicate floors 1 through 30 and 56 through 63. The Four Seasons will keep its floor numbers 35 through 39.[12]

Are you understanding all that? You can be in a forty-three-story building, but have a room on the sixty-third floor. How is such a thing possible?

I guess the regular numbering systems don't necessarily apply in Vegas.

But let's take a little closer look at things. Paddock was in a room on the thirty-second floor of the Mandalay Bay hotel, just two floors below the exclusive Four Seasons hotel. Might the shooting have had something to do with the Four Seasons hotel, located on floors thirty-five through thirty-nine (with a convenient separate elevator), which has escaped mention in most media accounts?

Would it be important to know who owns the Four Seasons hotel in Las Vegas? One needs only to consult a *Wall Street Journal* article from July 2017 to answer that question.

> A decade ago, two of the world's wealthiest men came together to buy Four Seasons Holdings Inc., home to some of the most expensive lodgings around.
>
> The deal was surprising, both for its lofty price tag, $3.8 billion, and for the unusual partnership, involving tech titan Bill Gates and Saudi Prince al Waleed bin Talal.
>
> The financial crisis soon pummeled the luxury hotel business. The partners then took to feuding, about matters ranging from helping fund new hotel developments to who should be chief executive. After a truce in 2013, they have been trying to whip their investment into shape.[13]

I previously mentioned my suspicion that a good deal of the attack on President Trump centered around what he wanted to do in the Middle East. Trump's tweet from 2015 about "Dopey Prince @Alwaleed_Talal" refers specifically to Saudi Prince al Waleed bin Talal, who at that time was widely expected to become the next King of Saudi Arabia.

By the way, I want to mention that George Soros had a short position on the MGM RESORTS INTERNATIONAL, that paid out if the stock went down. Since that hotel was where the massacre occurred, Soros got paid out well. You can still see the SEC filing for October 31st 2017:

https://www.sec.gov/Archives/edgar/data/1029160/000114036117031995/xslForm13F_X01/form13fInfoTable.xml

* * *

So, I started searching for what really happened in Las Vegas.

Saudi Arabia's immense oil and gas reserves gave them great power on the international stage. Like the pretty girl who doesn't need a good personality to be noticed, the Saudis didn't have to develop their economy. They could simply get away with pumping all that oil out of the ground.

But the fracking boom in the United States changed the situation. Suddenly, the United States had all the energy it needed. The global banking cartel put America at a disadvantage. By preventing America from developing its own oil, it required America to import it from places like Saudi Arabia. I believe that was done on purpose, in order to create a leash on the United States. If the United States doesn't go along with the cartel, they can ultimately yank the energy supply to the United States and cause an economic recession or even depression. In this control scheme, domestic production of oil of any nation threatens the cartel. The United States developing its own energy reserves and becoming self-sufficient is a direct threat to the global control structure.

The Saudis were faced with a dilemma. Their first response was to funnel some of their immense wealth into American environmental groups opposed to fracking. For many years, the Saudis had been generously donating to the Bush and Clinton families, effectively buying off both sides of the political aisle. They also made sure to spread their political largesse to many other politicians.

However, the fracking boom couldn't be stopped.

Then came Trump, an American businessman, who liked to make deals and didn't seem to be in love with endless wars.

King Salman (Salman bin Abdulaziz Al Saud) realized he wasn't going to be able to defeat American fracking and wanted to make a deal.[14] He wanted Trump's help in diversifying the assets of the country and setting up an industrial and manufacturing base separate from the oil industry.

Trump was open to the deal, but had some conditions of his own. The Kingdom had to stop interfering in American politics with their donations to Republicans, Democrats, and environmental groups. The Saudis also had to stop funding terrorist groups like ISIS, Hamas, and Hezbollah. In addition, the Saudis had to start opening up their oppressive society, starting with things like letting women drive.[15]

However, King Salman was eighty-one years old and his health wasn't good. A successor, or crown prince, who would continue the policy needed to be in place. In Saudi Arabia, being the crown prince wasn't necessarily a secure position. A crown prince was subservient to the wishes of the king and could easily be replaced. From January 2015 to April 2015, the crown prince was Muqrin bin Abdulaziz. From April 2015 until June 2017, the crown prince was Muhammed bin Nayef. In June 2017, shortly after Trump's visit to Saudi Arabia, King Salman named his thirty-two-year-old son, Mohammed bin Salman, referred to in the west as Crown Prince Salman, or alternately Crown Prince Mohammed bin Salman, and sometimes by his initials, MBS.

It was claimed that Mohammed bin Salman was present for the meeting in Riyadh between his father and Trump and was a strong supporter of liberalizing the Kingdom.[16] In late October 2017, Crown Prince Mohammed bin Salman gave a long interview to the British newspaper, *The Guardian*, in which he laid out his vision for Saudi Arabia:

> Saudi Arabia's crown prince, Mohammed bin Salman, has vowed to return the country to "moderate Islam" and asked for global support to transform the hardline kingdom into an open society that empowers citizens and lures investors.
>
> In an interview with the Guardian, the powerful heir to the Saudi throne said the ultra-conservative state had been "not normal" for the past 30 years, blaming rigid doctrines that have governed society in a reaction to the Iranian revolution, which successive leaders "didn't know how to deal with."[17]

The crown prince can only be looked upon as a reformer in the model of Mikhail Gorbachev from the former Soviet Union, or Nelson Mandela in South Africa. He was a young man in a hurry to modernize his wealthy, but conservative society with a history of religious intolerance.

> "What happened in the last 30 years is not Saudi Arabia. What happened in the last 30 years is not the Middle East. After the Iranian revolution in 1979, people wanted to copy this model in different countries, one of them is Saudi Arabia. We didn't know how to deal with it. And the problem spread all over the world. Now is the time to get rid of it."
>
> Earlier, Prince Mohammed had said: "We are simply reverting to what we followed—a moderate Islam open to the world and all religions. 70% of

These documents show how Google monitors the web for trends.

Google

Realtime Boost

felipeg@google.com

http://go/realtime-boost
http://go/realtime-boost-demo
http://go/realtime-boost-howto

Realtime Boost - Indexing

Freshdocs-instant Crawl

Dominicans of Haitian descent tired of bias at the ballot box

Aftermath of a crisis

Pubsub CDOC

Indexer

Time
Unigram
Location
Entity
Salient Terms
...

Hivemind - Tokenized - Muppet Index

```
Global ExtendedDocId: 10529037701726682924::0::0
LocalDocId: 1029292, Global ExtendedDocId: 10529037701726
Section body (Pos 131749376 -> 131749492)
          TokenPos    TokenId                token   a
         131749376       2983 [  <global_delimiter>]
         131749378    6362733 [        rthh=811207]
         131749379    4023159 [        rthh=811208]
         131749380       3138 [            rtl=eng]
         131749381      77581 [         rtw=voting]
         131749382       4097 [            rtw=cnn]
         131749383    4025558 [     rtw=dominicans]
         131749384     140670 [        rtw=haitian]
         131749385     100855 [        rtw=descent]
         131749386     108206 [          rtw=tired]
         131749387     240848 [           rtw=bias]
         131749388      86357 [        rtw=mariano]
         131749389      86358 [       rtw=castillo]
         131749390      69870 [            rtw=img]
         131749391       4424 [        rtw=updated]
         131749395       6157 [            rtw=edt]
         131749398       9657 [          rtw=month]
         131749399       4140 [           rtw=away]
         131749400       8144 [   rtw=presidential]
```

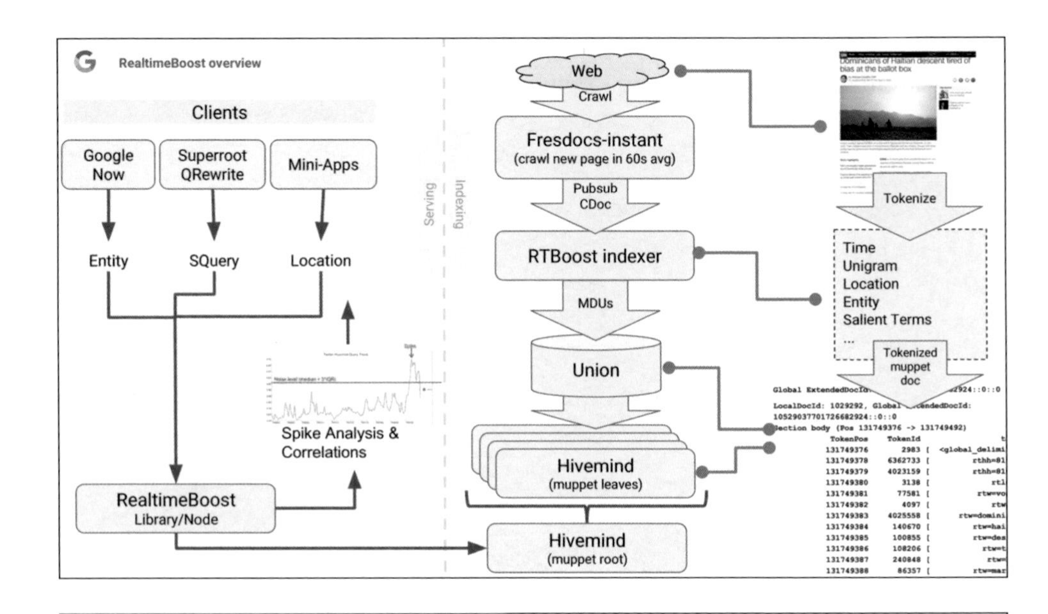

Realtime Hivemind Scoring (go/hivemind-scoring)

$$LiftScore = \frac{P(event \mid query)}{P(event)}$$

Event is the time of the document publication, quantized to 30 minutes e.g. time=794237.

LiftScore can be interpreted as how much more likely to see document matching the query at the given time, comparing to seeing random document at the given time.

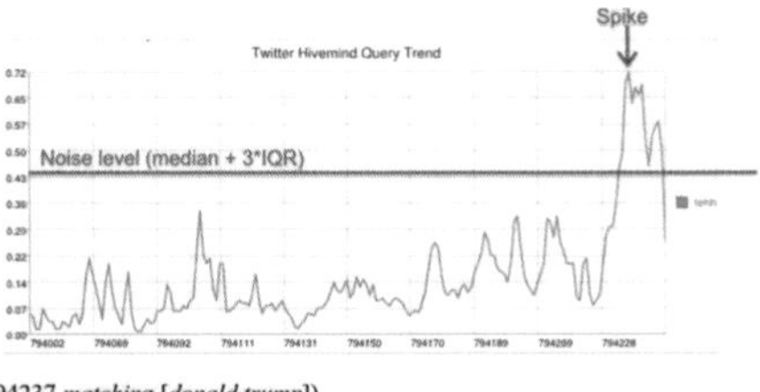

$$LiftScore = \frac{P(time{=}794237 \mid [donald\ trump])}{P(time{=}794237)}$$

$$P(time = 794237) = \frac{N(docs\ at\ time{=}794237)}{N(docs\ total)}$$

$$P(time = 794237 | [donald\ trump]) = \frac{N(docs\ at\ time{=}794237\ matching\ [donald\ trump])}{N(docs\ matching\ [donald\ trump])}$$

Google

No birds were hurt in the making of Realtime Boost signal

Google's obsession over a sitting president is amazing. These slides show how Google redesigned their own news systems to hurt the sitting president.

Realtime Boost Events

- **Realtime** - Realtime detection of news-events - Under 5 minutes after event started.
- **Event Understanding**
  - Build Correlations in multidimensional space
  - Temporal locality
  - Temporal topicality
    - Entities / Salient-Terms / Summary / Label / Queries / Questions / Videos, etc...
    - Event Popularity (CV / NB) and importance
- **Query Understanding** - Search currently trending Events for the Query
  - Also for Query-less, using list of Entities / Salient Terms / Google-Now profile
- **Statefulness** - Understand Story updates and checkpoints
- **Fine Grained and precise in time**

Google

Confidential + Proprietary

Trump Fires Comey - News Cluster

(news cluster link)

The Original Cluster has been transformed into everything Comey/Russia/Trump

Mishmash of new developments:

- **Comey Memo / Russia probe**
- **FBI next pick**
- **Russia Leaks / Classified info**

News Cluster is active for 8 days

- **18,549 articles so far**

Realtime Coverage

Chaffetz: I'm ready to subpoena for Comey memo

If the president isn't hiding anything, what is the need for his secrecy?

Trump: FBI job hunt on fast track

Americans Deserve To Know About Russian US Election Meddling | Senator Casey

Most important editorial of the weekend: It's time to consider impeachment

Democrats continue to pressure Trump over FBI boss' axing

In Depth

Key Moments of Trump-Comey Clash and What's Come After

Will Firing Comey Cost Trump?

Trump, GOP control FBI chief pick despite Democrats' calls

Yes, Trump Thinks He Can Defeat The Russia Probe

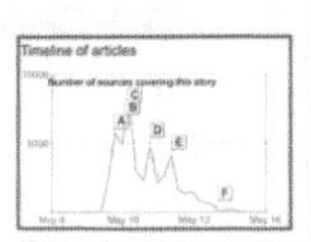

Google

Trump Fires Comey - RTB Event

(query)

- **Original Event is only about Trump firing Comey**
  - **Precise in time and in topic**
  - **Trump leaking confidential information to Russia is a different story**
- **Timeline brings new developments**
  - **Still very related to Trump firing Comey**

Trump Fires Fbi Director Comey
President Donald Trump Fires Fbi Director James Comey
President Donald Trump abruptly fired FBI Director James Comey Tuesday, dramatically ousting the nation's top law enforcement official in the midst of an FBI investigation into whether Trump's campaign had ties to Russia's meddling in the election that sent him to the White House.
Frequently Asked:
Why did trump fire the fbi director?
Why did james comey get fired?
Why was james comey fired?

Story Timeline

Trump pulling out of Paris agreement

(query)

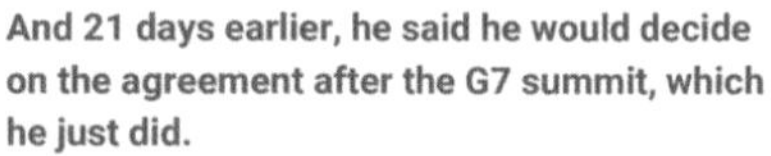

Story Timeline

4 days 4 hours ago
G 7 Source Us Gets More Time On In Out Climate Pact Answer

And 21 days earlier, he said he would decide on the agreement after the G7 summit, which he just did.
(also from the timeline)

21 days 21 hours ago
Paris Climate Accord
Trump Will Make Decision On Paris Climate Pact After G7 Summit
On Tuesday, the White House has said that US President Donald Trump will not make a decision on whether the US should remain in the Paris climate agreement until after he returns from the G7 summit.
Inspect Event Data - Search Similar Events
Story Similarity Score: 1760.5

Trump tweets covfefe. The *New York Times* says it doesn't exist and in lockstep Google deletes the word from its Arabic translation dictionary.

Donald J. Trump
@realDonaldTrump

Despite the constant negative press covfefe

RETWEETS 22,602 LIKES 27,470

9:06 PM - 30 May 2017

Donald J. Trump
@realDonaldTrump

Who can figure out the true meaning of "covfefe" ??? Enjoy!

3:09 AM - 31 May 2017

The New York Times

By Liam Stack

June 1, 2017

The internet is full of confident people who do not know what they are talking about. The latest example: a conviction spreading in right-leaning social media communities that a garbled tweet by the president — he wrote "covfefe" — was not a late-night typo but was instead Mr. Trump sending a message to the world in Arabic.

'covfefe' Translate Easter egg

Goal

We currently translate the query "cov fe'fe" from Arabic to English into "I will stand up". This created some confusion this week as users tried to translate Donald Trump's tweet from Wednesday night which had the word "covfefe" in it. Since the word has no real meaning, we want to do an Easter egg that translates "cov fe'fe" and "covfefe" into "¯_(ツ)_/¯" on Translate properties.

See article: https://www.nytimes.com/2017/06/01/us/politics/covfefe-trump-arabic.html

62301306 Add "covfefe" Easter egg Comment revision history is now live! go/comment-history **Dismiss**

Rosetta (Google Translate) Web / TWS

tiberius@google.com <tiberius@google.com> #1 Jun 2, 2017 02:50PM

Reported Issue, Assigned to tiberius@google.com

The words "covfefe" and "Cov fe'fe" (normalized for things like capitalization) will translate to "¯_(ツ)_/¯" from any language to any other language. We have a demo up and running: https://translate-qa2.sandbox.google.com/#eu/en/cov%20fe'fe, but there are some issues with auto-detect because it gets auto-detected to Arabic and we do some pre-processing on the "Cov fe'fe" phrase. Therefore we will find out what that phrase gets changed to and also intercept that 3rd string.

All of this will be behind a Mendel flag. To ramp it up and down you will have to go to http://mendel/ and change the percentage from 0% to 100% (or the other way around).

Macduff Hughes <macduff@google.com> #2 Jun 2, 2017 02:55PM

+cc coliny, qge (oncall now and for the weekend)

tiberius@google.com <tiberius@google.com> #3 Jun 2, 2017 03:29PM

Macduff & Julie, I discovered that it's hard to wrap the Easter egg in a Mendel flag without substantial refactoring. Still looking, but this could be a launch blocker. We don't have time to refactor today, and without Mendel rolling back would take hours of engineering work on a weekend.

tiberius@google.com <tiberius@google.com> #4 Jun 2, 2017 03:44PM

The string the "Cov fe'fe" gets rewritten to in Arabic (the auto-detected language) is "سوف ققف", so this is the variant we'd have to support. I will send out the CL that adds this Easter egg shortly. It does not have Mendel support; it would just be a 100% launch to production. The "Funniest Joke in the World" Easter egg also wasn't controlled by Mendel. I will happily push this change if it's approved, but I make no guarantees about being available for it during the weekend.

Fake News-letter

BY gTech Search + Content
POC: [redacted]@

NOVEMBER 27, 2017 | INTERNAL ONLY

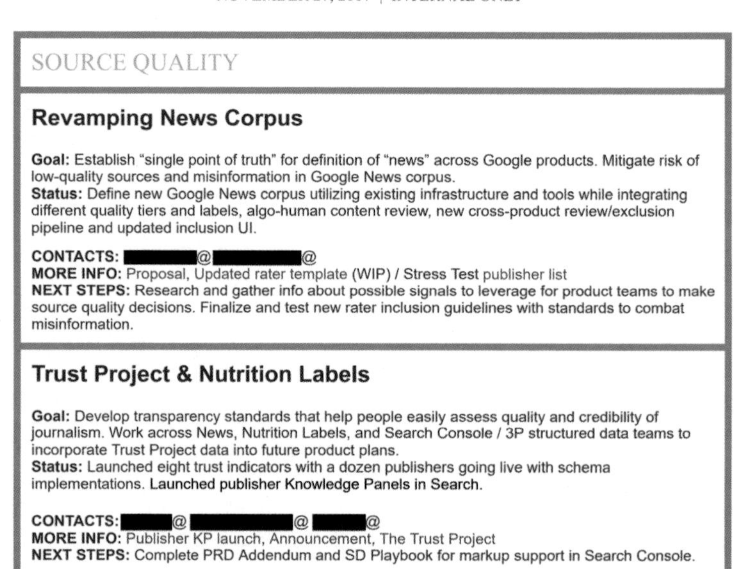

SOURCE QUALITY

Revamping News Corpus

Goal: Establish "single point of truth" for definition of "news" across Google products. Mitigate risk of low-quality sources and misinformation in Google News corpus.
Status: Define new Google News corpus utilizing existing infrastructure and tools while integrating different quality tiers and labels, algo-human content review, new cross-product review/exclusion pipeline and updated inclusion UI.

CONTACTS: [redacted]@ [redacted]@
MORE INFO: Proposal, Updated rater template (WIP) / Stress Test publisher list
NEXT STEPS: Research and gather info about possible signals to leverage for product teams to make source quality decisions. Finalize and test new rater inclusion guidelines with standards to combat misinformation.

Trust Project & Nutrition Labels

Goal: Develop transparency standards that help people easily assess quality and credibility of journalism. Work across News, Nutrition Labels, and Search Console / 3P structured data teams to incorporate Trust Project data into future product plans.
Status: Launched eight trust indicators with a dozen publishers going live with schema implementations. Launched publisher Knowledge Panels in Search.

CONTACTS: [redacted]@ [redacted]@ [redacted]@
MORE INFO: Publisher KP launch, Announcement, The Trust Project
NEXT STEPS: Complete PRD Addendum and SD Playbook for markup support in Search Console.

This internal document shows that Google was consolidating news across the whole company into a central executive structure to determine what "Fake News" was.

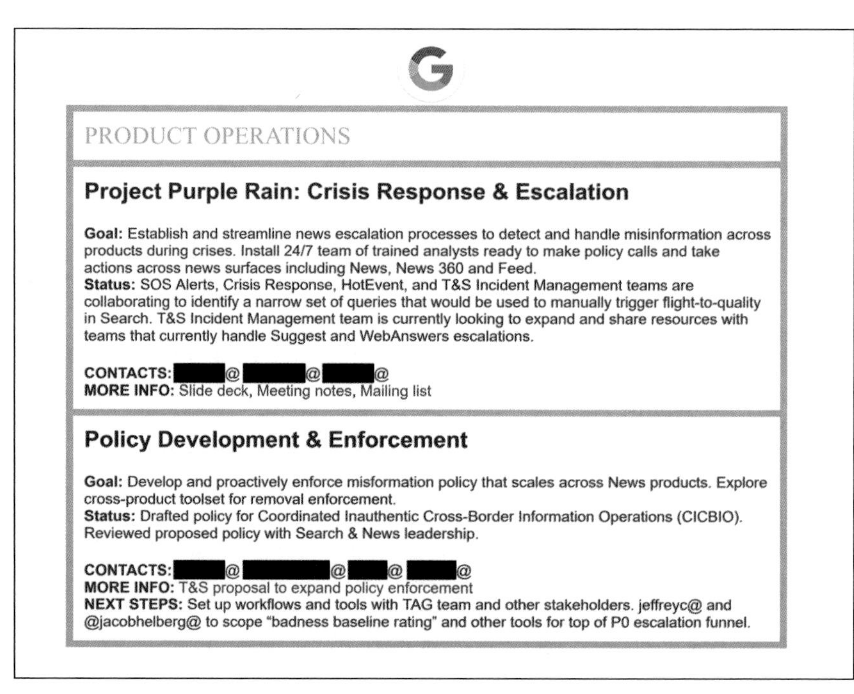

PRODUCT OPERATIONS

Project Purple Rain: Crisis Response & Escalation

Goal: Establish and streamline news escalation processes to detect and handle misinformation across products during crises. Install 24/7 team of trained analysts ready to make policy calls and take actions across news surfaces including News, News 360 and Feed.
Status: SOS Alerts, Crisis Response, HotEvent, and T&S Incident Management teams are collaborating to identify a narrow set of queries that would be used to manually trigger flight-to-quality in Search. T&S Incident Management team is currently looking to expand and share resources with teams that currently handle Suggest and WebAnswers escalations.

CONTACTS: [redacted]@ [redacted]@ [redacted]@
MORE INFO: Slide deck, Meeting notes, Mailing list

Policy Development & Enforcement

Goal: Develop and proactively enforce misformation policy that scales across News products. Explore cross-product toolset for removal enforcement.
Status: Drafted policy for Coordinated Inauthentic Cross-Border Information Operations (CICBIO). Reviewed proposed policy with Search & News leadership.

CONTACTS: [redacted]@ [redacted]@ [redacted]@ [redacted]@
MORE INFO: T&S proposal to expand policy enforcement
NEXT STEPS: Set up workflows and tools with TAG team and other stakeholders. jeffreyc@ and @jacobhelberg@ to scope "badness baseline rating" and other tools for top of P0 escalation funnel.

"Anyone want to hold their nose and look through Breitbart.com for hate speech?"

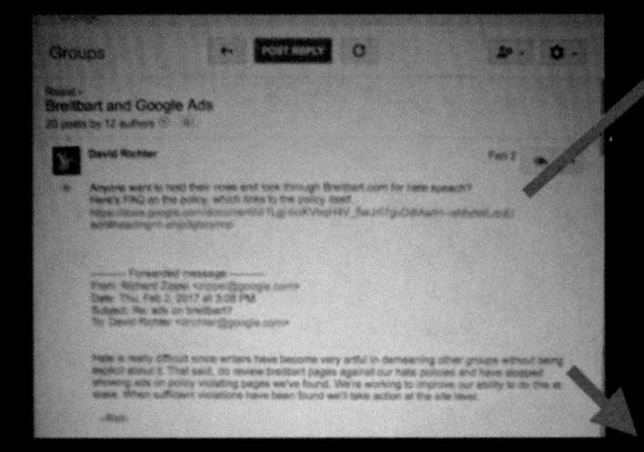

"Hate is really difficult since writers have become very artful in demeaning other groups without being explicit about it. That said, do review breitbart pages against our hate policies and [we] have stopped showing ads on policy violating pages we've found. We're working to improve our ability to do this at scale. When sufficient violations have been found we'll take action at the site level."

Internal communications of employees plotting to sabotage *Breitbart's* revenue stream.

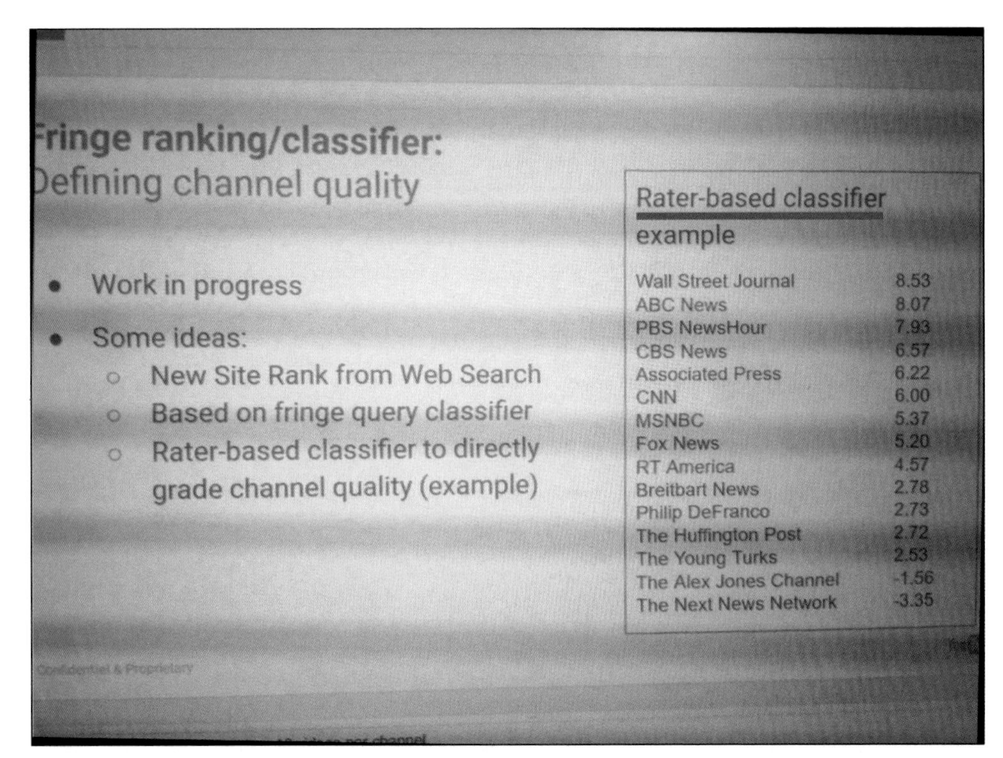

Executive communications showing the different ranking scores of news outlets.

```
# YoutubeControversialTwiddler Query Blacklist.
#
# The queries in this file will force the controversial twiddler to be
# triggered.
#
# This file will be parsed by superroot::StringLookupModule. The data format is
# plain text. Each line contains one query, without leading/trailing/duplicated
# whitespaces. Each term is delimited by a single space.
64 old stephen paddock
64 stephen paddock
64 steven paddock
64 years old stephen paddock
64 yr old stephen paddock
about stephen paddock
adress trump las vegas
anti trump paddock
anti trump stephen paddock
anti trump steve paddock
anti trump vegas
anti-trump paddock
atack las vegas oktober 2017 hoax
atack lasvegas
ataq las vegas
ataque lavegass
atephen paddock
attack a las vegas
attack from the left las vegas
attack in las vegas
```

This YouTube query blacklist found inside of Google suppressed searches for sensitive topics. Roughly twenty out of forty pages were dedicated to the Las Vegas Massacre.

```
abortion increases the risk of ectopic pregancy by 50
abortion increases risk of mental health problems
abortion increases risk of premature birth
how to have an abortion without going to the doctors
how to have an abortion without a doctor
how to have an abortion at home yahoo
how to have an abortion using coca cola
Unrestricted abortion
Abortion up to 12 weeks
The 8th Amendment of the Constitution of Ireland
Repeal the 8th
Repeal the 8th / abortion Fine Gael
Repeal the 8th / abortion Fianna Fail
Repeal the 8th / abortion Sinn Fein
Repeal the 8th / abortion Labour
Repeal the 8th / abortion Green Party
Repeal the 8th / abortion Varadkar
Repeal the 8th / abortion Coveney
Repeal the 8th / abortion Harris/Simon Harris
Repeal the 8th / abortion Martin
Repeal the 8th / abortion Mary Lou McDonald
Repeal the 8th / abortion Howlin
Repeal the 8th / abortion Higgins/President Higgins
Repeal the 8th / abortion Cardinal Daly
Abortion and rape
Abortion and incest
Abortion and Down Syndrome
Woman's right to choose
Child murder
Fatal foetal abnormalities
```

YouTube decided to limit the Irish from searching for their own constitutional amendment during an election.

```
adl
anti defamation league
bnai brith
gun owners of america
heritage foundation
naacp
nra
planned parenthood
splc
gop train crash
gop train crash assassination
gop train crash assassination attempt
gop train crash attempted assassination
gop train crash attempted killing
gop train crash theory
attempted assassination of gop congress members
intentional gop train crash
gop train
train crash
911
```

```
crisis actors florida
crisis actors florida shooting
crisis actors school shooting
crisis actors parkland
crisis actors conspiracy
crisis actor florida
florida hoax
false flag florida school shooting
false flag florida
false flag florida shooting
black lives matter
david hogg the actor
david hogg fraud
david hogg practicing lines
david hogg rehearsal
hogg crisis actor
hogg actor
shooting hoax
florida high school shooting hoax
false flags
austin
austin texas
austin bombing
austin explosion
austin bomb
austin news
bombing austin
austin bombings
austin bombs
```

Video's showing David Hogg forgetting his lines were censored on Google.

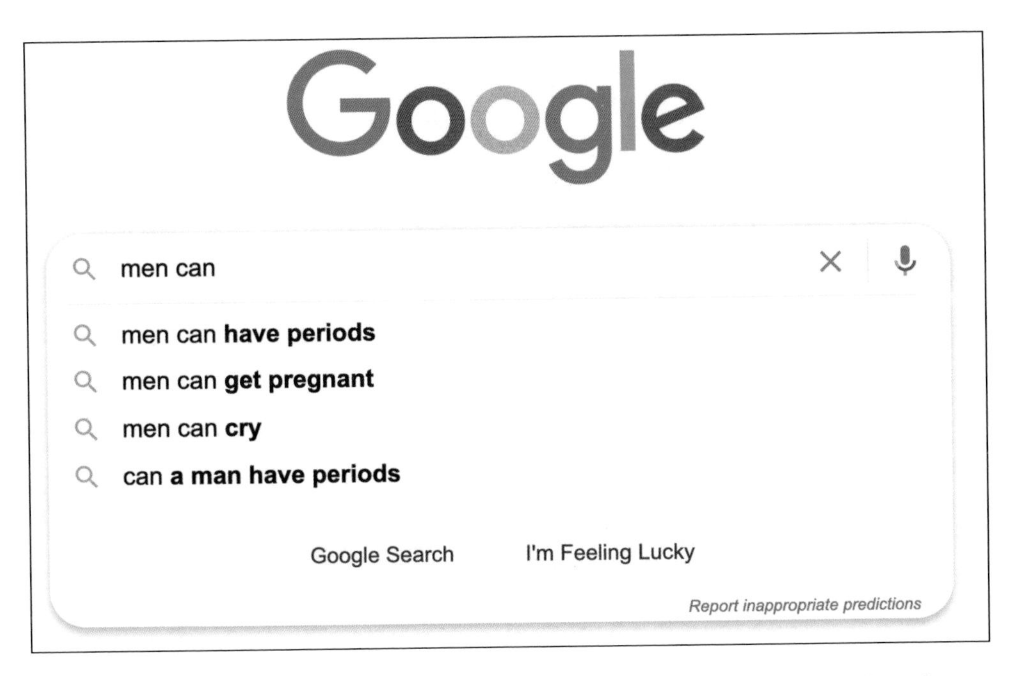

Google showing their "anti-science" side by asserting men can have periods and get pregnant.

That's me at age two with my dog, Pete.

Ever since fifth grade when I got my hands on my first Mac, I've been addicted to computers.

Like many of my generation, I became interested in the Occupy Wall Street movement, protesting the big bank bailouts of 2009.

Since college I've loved motorcycles. And yes, sometimes I'm surprised I'm alive.

I became fascinated with using technology to make jewelry and improve the motorcycle riding experience.

Coming out with my hands up after Google sent the police after me.

Revealing myself on Project Veritas was both terrifying and a liberation of my soul.

> Saudis are younger than 30. Honestly, we won't waste 30 years of our life combating extremist thoughts. We will destroy them now and immediately."[18]

It's probably puzzling to the average person to read those words of the crown prince of Saudi Arabia. The most powerful person in the Middle East (arguably) taking such a strong stance against radicalism and advocating for a more tolerant and inclusive Islam?

His persuasion pitch was perfect.

Why didn't this interview receive wall to wall coverage in Western media?

I return once again to the view of the alternative narrative. Not because I can tell you it's true, but because it provides a different perspective.

In the world of the alternative narrative, Western intelligence services, like the CIA, are manipulating events behind the scenes in order to keep tensions high enough to justify massive arms deals for defense contractors, but just shy of major conflict between the nuclear superpowers. For many years, that was a talking point of the anti-war left, who highlighted the lies of the Vietnam War, and among a later group of activists who claimed the Bush administration lied us into a war with Iraq by phony claims of weapons of mass destruction in that country.

Western intelligence is supposedly active in places like North Korea and Iran, stoking tensions and paranoia among the leadership. In other countries, like Afghanistan, the intelligence services are supposedly profiting off the heroin trade. In Saudi Arabia, the intelligence services are making money off of human trafficking in the sex trade, providing prostitutes to Europe, the Middle East, and Asia. Saudi Arabia, with its great wealth, and closed society, provides such a fertile ground for the intelligence services it's claimed they refer to the country as "Wonderland." In other words, the "bad princes" of Saudi Arabia are likely working with the intelligence services of the West to keep things just the same.

So, if you find yourself wondering why you didn't realize the crown prince of Saudi Arabia was embarking on an unprecedented loosening of authoritarianism in the Middle East, the answer might be as simple as the mainstream media didn't want you to know.

* * *

And now we return to the events of October 1, 2017.

At 9:59 p.m. somebody shot through the door of Stephen Paddock's room at a security guard. For about ten minutes starting at approximately 10:05 p.m. more than eleven hundred rounds of ammunition were fired at the crowd below. At 11:20 p.m. security personnel finally breached the door, entered the room, and found Stephen Paddock dead.

This took place on the thirty-second floor of the Mandalay Bay Hotel, just two floors below the Four Seasons Hotel, owned by Bill Gates and Prince Alwaleed bin Talal.

According to some accounts, the Four Seasons Hotel was completely booked that night by a Saudi Arabian delegation, including Crown Prince Mohammed bin Salman.[19] (According to some accounts, there was a Saudi security summit booked in Vegas that weekend. Perfect cover for a lot of rich Saudis to be in the city.) The plan by the king, the crown prince, and a number of "good princes" to reform their country had created some serious opposition.

The "bad princes," of which Prince Alwaleed was a member, had a plan. And to make it work, they needed to involve the American authorities. Saudi intelligence would tell the Americans, presumably the FBI and/or CIA, that terrorists were looking to operate in the United States, and they needed to set up a "sting" operation. However, the "terrorist" team would actually be an assassination squad to kill the crown prince.

The King of Saudi Arabia was in Russia at that time, and the assumption was they'd find a way to kill him on the way back, or shortly after he returned to the Kingdom. With the crown prince and the king dead, the "bad princes" would maneuver the former crown prince, Muqrin bin Abdulaziz, to be the new king. Presumably because the most recent crown prince, Muhammed bin Nayef, had led the negotiations with the Americans which resulted in the large arms deal Trump brought with him, as well as the agreements to stop interfering in American politics and funding radical terrorist groups.

But the terrorists/assassination squad couldn't fly into the United States with weapons. They'd need to procure their weapons in Las Vegas.

Enter Stephen Paddock, our "mystery man." The allegation is that Paddock was a contract employee for either the FBI or CIA, the perfect figure who would not arouse suspicion.

The assassination team would probably consist of around twenty individuals, since they'd probably be taking on a force of somewhere around thirty of the crown prince's bodyguards. But they'd have the element of surprise, right? Who'd be expecting an attack from the thirty-second floor of the Mandalay Bay? Sounds crazy, right? The team meeting Paddock would be small, maybe just a few men, they'd inspect the weapons, make sure they were good for their purpose, kill Paddock, and then radio the rest of the assassination squad to come on up.

All of this might have worked, except for one mistake.

If you're a young, wealthy Saudi prince and you're in Las Vegas, do you want to stay in your hotel room? Saudi princes have been known to put on a disguise, take just a few bodyguards who could easily be mistaken as just a few guys, and head out to the Vegas strip, maybe to the Tropicana Hotel.

At the Tropicana Hotel, a few members of the assassination squad started a gunfight in an attempt to take out the crown prince, but were driven back. (Video footage of what's alleged to be the crown prince at the Tropicana, as well as people shot at the Tropicana would later appear on YouTube.[20])

How does it go from there? We can only speculate.

Perhaps the crown prince's security detail now knew there was a threat, but it wasn't clear whether there were additional gunmen.

Trouble had already started at the Mandalay Bay Hotel because a security guard found that access to the thirty-second floor was barricaded, and he went to investigate, forcing somebody inside the room to shoot through the door.

The gunmen inside Paddock's room needed to start a diversion to see if they can make an escape. There are unfriendly Saudi security services above them, and American security forces advancing from below.

They kill Stephen Paddock, start loading guns, break a window, and begin firing into the crowd below. One of the gunmen goes to another window and starts firing at some large aviation fuel tanks at the nearby airport. An explosion of these tanks would surely be a good diversion, giving them a chance to escape. But the fuel tanks don't explode.

The assassins in Paddock's room, after ten minutes of mayhem, realize it's hopeless, and kill themselves.

Meanwhile, the crown prince and his bodyguards have fought off the members of the assassination squad, and escape in a helicopter from the roof of the Mandalay Bay. (Video of the helicopter escape also circulates on YouTube.[21]) The FBI is now assisting the crown prince to escape and turn to the gunmen in Paddock's room.

FBI members join the SWAT team, burst into the room, and find Paddock dead, and the two Saudi assassins. They remove the bodies of the assassins, take pictures, and "release" them on 4Chan, a website known to be used by those who question the mainstream narrative.

The clean-up operation was now in full swing.

* * *

The implications of this shooting are enormous.

If the public believed elements of the Saudi (or other) security services carried out the greatest mass shooting of Americans in modern times, it's difficult to imagine how we would avoid yet another war in the Middle East.

If the public believed American law enforcement (either the CIA or FBI) was duped into providing the weapons to these killers, these agencies might be splintered into a thousand pieces to satisfy an angry public.

If the Arab street believed that some of their radical brethren again took their holy war to the United States and almost took out the crown prince of Saudi Arabia, who knows how many would dedicate themselves to similar martyrdom?

* * *

Was there an attempt on the life of the King of Saudi Arabia on October 4, 2017? An article in the Israeli newspaper *Haaretz* suggested that possibility.

> Yemen's Iran-backed Houthi rebels on Tuesday said they fired a ballistic missile targeting the al-Yamamah royal palace in the Saudi capital of Riyadh, Houthi-affiliated TV al-Masirah reported.
>
> Saudi air defenses intercepted the missile, Saudi-owned channel al-Arabiya reported in a news flash quoting a Saudi-led military coalition. Coalition forces carried out airstrikes on Houthi positions in Yemen's southern Sanaa in response, al-Arabiya reported . . .
>
> Saudi leaders were reportedly meeting at the palace at the time of the fire, Houthi media reported. There were no immediate reports of any damage as a result of the intercept, state TV reported.[22]

It's not an understatement in the Middle East to say anything is possible. But let's go over the claims which have a strong basis in established fact. Rebels in Yemen are backed by Iran, and their opposition is backed by Saudi

Arabia. Saudi Arabia doesn't want a government *friendly* to Iran on its border. How difficult would it have been for the Houthi rebels to know that Saudi leaders were meeting in that particular palace and fire a ballistic missile at them? Maybe that was in the range of Yemeni, Houthi, or Iranian intelligence assets. Or maybe there was an assist by some of the so-called "bad princes" in Saudi Arabia. How might we figure it out, when by its very nature, warfare relies on deception and concealment?

Maybe we look at what happened afterward.

When America was attacked on September 11, 2001, the country united in response to the threat.

But in Saudi Arabia, after Las Vegas, and after the ballistic missile attack that was targeted on a royal palace where Saudi leaders were supposedly meeting, a virtual civil war broke out in the Kingdom.

How do you cover up a counter-coup in a very secretive society?

Maybe you call it "an anti-corruption purge?" This is how it was reported in the pages of the *New York Times*:

> Saudi Arabia announced the arrest on Saturday night of the prominent billionaire investor Prince Alwaleed bin Talal, plus at least 10 other princes, four ministers and tens of former ministers.
>
> The announcement of the arrests was made over Al-Arabiya, the Saudi-owned satellite network whose broadcasts are officially approved. Prince Alwaleed's arrest is sure to send shockwaves both through the kingdom and the world's major financial centers.
>
> He controls the investment firm Kingdom Holding and is one of the world's richest men, owning or having owned major stakes in 21st Century Fox, Citigroup, Apple, Twitter, and many other well-known companies. The prince also controls satellite television networks watched across the Arab world.[23]

Was it all merely a coincidence that these arrests happened so soon after the Las Vegas massacre, which supposedly had NO connection to Saudi Arabia? If one considers Saudi Arabia as a big, rich group of princes connected by blood and marriage, closed off in many ways from the rest of the world, it makes a good deal of sense. If the crown prince was going to Vegas, then of course he'd have to stay at the Four Seasons Hotel. To do otherwise would be an insult to Saudi Arabia.

The Kingdom is proud of Prince Alwaleed, who likes to style himself as the "Warren Buffet of Arabia"[24] after the very successful American stock picker. Alwaleed has also been at it for a very long time. A profile from the *New York Times* in 1999 opened this way:

> He calls them his 100 wives and honors each with a flag tacked to his office wall. Citigroup. Saks Fifth Avenue. Four Seasons Hotel. Apple Computer. Movenpick. Saatchi & Saatchi. Daewoo. Donna Karen International. Trans World Airlines. The News Corporation. Planet Hollywood. Hyundai Motor.
>
> It is an extraordinary group, all the more so because this one man, Prince Walid bin Talal of Saudi Arabia, owns at least 5 percent of them—the core of a fortune that he says is now worth $14.2 billion dollars.
>
> That represents a tenfold increase from just 10 years ago, the fruit of an investing binge that has won the Prince, now 44, renown as one of the world's sharpest stock pickers and catapulted him to a place behind William H. Gates of Microsoft . . .[25]

Talal's business interests make him the perfect person that plotters would need to move all the pieces necessary for a successful assassination plot. *Need to find a time when the crown prince will be vulnerable? How about when he's in Las Vegas? I can get him to stay in my hotel.*

Maybe that's what happened.

Or maybe there's another reason the crown prince arrested the "Warren Buffet of Arabia" at the start of his anti-corruption drive.

> The king had decreed the creation of a powerful new anti-corruption committee, headed by the crown prince, only hours before the committee ordered the arrests.
>
> Al Arabiya said that the anticorruption committee has the right to investigate, arrest, ban from travel, or freeze the assets of anyone it deems corrupt.
>
> The Ritz Carlton hotel in Riyadh, the de facto royal hotel, was evacuated on Saturday, stirring rumors it would be used to house detained royals. The airport for private planes was closed, arousing speculation that the crown prince was seeking to block rich businessmen from fleeing before more arrests.[26]

This was a committee that didn't waste its time. Within a few hours of forming, it had arrested some of the most powerful men in the country. The powers the committee held were breathtaking. They could investigate, arrest, ban, or freeze the assets of anybody they wanted. They'd closed down the Ritz Carlton hotel and were planning to use it as a temporary jail. (Only in Saudi Arabia!) The private airport used by the elite was shut down. The rats were trapped.

Was it part of a vendetta which involved President Trump? The *New York Times* article on the arrest seemed to suggest that possibility, writing of Prince Alwaleed:

> He has also recently sparred with President Donald J. Trump. The prince was part of a group of investors who bought control of the Plaza Hotel in New York from Mr. Trump, and he also bought an expensive yacht from him as well. But in a twitter message in 2015, the prince called Mr. Trump "a disgrace not only to the GOP but to all America. Withdraw from the U.S. presidential race as you will never win."[27]

It certainly seemed as if far more was going on beneath the surface than was being reported in the mainstream media. What had stirred Saudi Arabia into such a state that it was taking such drastic actions? One can only assume it was a crisis which threatened the very foundation of the country.

It wasn't some run of the mill corruption investigation.

Something much larger was taking place.

The world was simply being kept in the dark as to what was really happening.

* * *

Can we find any evidence of what happened during this "anti-corruption" campaign to the family of former Crown Prince Muqrin? There were suggestions that if Muqrin became king, then his son, Prince Mansour bin Muqrin, would become the new crown prince. Unfortunately, on October 5, 2017 Prince Mansour died in a helicopter crash, according to an *Al Jazeera* article:

> A helicopter carrying several Saudi officials, including a high-ranking prince, has crashed in the kingdom's southwest near the border with Yemen.
>
> Prince Mansour bin Muqrin Al Saud, deputy governor of Asir province, was on a tour of local projects west of the city of Abha when the crash happened on Sunday evening . . .
>
> The son of the former Crown Prince Muqrin bin Abdulaziz, Prince Mansour bin Muqrin was appointed deputy governor of Asir province, which borders Yemen, earlier this year.[28]

The coincidences in Saudi Arabia seem to keep piling up. Prince Mansour had been appointed deputy governor of a province "which borders Yemen," which is exactly where the troubles with Iran, and that pesky ballistic missile, came from. Was he making a run for Yemen, and possibly Iran, to escape the anti-corruption campaign?

On October 8, 2017, the *Middle East Monitor*, a not for profit press monitoring operation founded in 2009, published allegations that Prince Mansour's helicopter had been brought down by Saudi security services.

> The helicopter carrying Saudi Prince Mansour bin Muqrin and seven other people was deliberately targeted by state forces because it is believed Bin Muqrin opposed Crown Prince Mohammed Bin Salman's succession to the throne, informed sources told the New Khaleej.
>
> According to the sources, Bin Muqrin had recently sent a letter to some 1,000 princes urging them to step away from support for Bin Salman's succession to the throne, pointing out that the youth must take action . . .
>
> The incident came less than a day after authorities arrested 11 princes and 40 former ministers and senior officials, most notably Head of the National Guard Prince Miteb Bin Abdullah and billionaire Prince Al-Waleed Bin Talal for alleged involvement in corruption.[29]

In a traditional society like Saudi Arabia, what does it mean when a leading prince sends a letter to more than a thousand princes, urging them to oppose the succession of the crown prince? It's almost as unprecedented as Democrats in the United States urging the members of the Electoral College to change their vote from Donald Trump to Hillary Clinton in what became known as the "faithless elector" affair.

And on exactly what basis was Prince Muqrin opposing the succession of Mohammed Bin Salman? Was it because he wanted to defang radical Islamist groups?

A narrative was beginning to form.

On October 1, 2017, an attempt was made on the life of Crown Prince Mohammed Bin Salman in Las Vegas, avoided only because the crown prince decided he wanted to go gambling at the Tropicana Hotel.

On October 3, 2017, the King of Saudi Arabia called a special meeting to create an anti-corruption committee, placing the crown prince at the head of it, providing it with wide-ranging powers to investigate, detain, and question individuals. The "Arabian Warren Buffet," Prince Al-Waleed, is arrested, as well as many princes, former government ministers, and the former head of the National Police. The Ritz Carlton was being prepared as a jail for those arrested and the private airport used by the wealthy was closed.

On October 4, 2017, a ballistic missile is fired from Yemen at the Royal Palace in Riyadh, which was in the middle of an important government meeting.

On October 5, 2017, Prince Muqrin, who was supposedly going to take over as the new crown prince after the coup, died in a suspicious helicopter crash, which some claim was brought down by Saudi security services.

From my read of the situation, a counter-coup had been orchestrated in Saudi Arabia, and people were trying to cover it up as merely an "anti-corruption probe." Just one authoritarian ruler going after his perceived enemies. It happens all the time in those kinds of countries.

Nothing to see here, people; move along.

* * *

What did a mass shooting in Las Vegas and trouble in the Middle East have to do with Google?

Remember how Google wanted to do "real-time" interventions on breaking news stories? Well, the Las Vegas massacre was the test case for their new system of information control, and I was able to find that document.[30] It was a dragnet of news stories, eventually capturing 427 articles. It had been widely reported that Prince Alwaleed was heavily invested in tech companies and hated Donald Trump. Were the tech companies part of the Saudi cover-up? Or was it simply a coincidence that the attempt to deal with "fake news" took place at the same time as the Las Vegas Massacre and the Saudi Arabian "anti-corruption" campaign?

For a flavor of the articles Google didn't want you to see, I'll give you the first five stories they de-listed.

From *CNN*, an article titled "Las Vegas Attack the Deadliest US Mass Shooting."[31]

From *ABC News*, an article titled "Las Vegas Attack Deadliest Shooting in Modern US History."[32]

From the *Express* in England, "Las Vegas Shooting, Mandalay Bay Attack Emergency Phone Number Hotline."[33]

Another article from the *Express* with the title, "Las Vegas Shooting—What Happened in Mandalay Bay Terror: Latest Attack News Update."[34]

A third article from the *Express* rounded out the first five: "Las Vegas Shooting Updates: Latest News: Shooter Death Toll of Victims."[35]

As I went through the document there were a couple stories they'd de-listed which caught my attention.

De-listed article 236 was from *TMZ*, with the title, "Mass Shooting at Mandalay Bay Casino in Vegas: Multiple Shooters Suspected."[36]

De-listed article 311 was from a discussion group called Longroom: "Mass Shooting at Mandalay Bay Casino in Vegas: Multiple Shooters Suspected."[37]

De-listed article 327 was from *Patch*, with the title "Las Vegas Mass Shooting Reports: Other Officers Down."[38] Could these "other officers" have been shot in the course of the alleged attack on the crown prince at the Tropicana Hotel?

De-listed article 391 was from *Follow News*: "Horror Show in Vegas: Eyewitnesses Recount Chaos of Mandalay Bay Shooting."[39]

Google had deployed their real-time censoring of breaking news stories to control the things the world was likely to see.

If you didn't know there were other narratives, and conflicting accounts, it would never even cross your mind. The public was being brainwashed in a manner so subtle, you'd never know it was happening.

But I could see it all.

And because I saw it, I had a duty to act, regardless of the consequences to myself.

What do you do when you realize you've been working for the bad guys?

Luckily, the bad guys had been telling everybody for more than six months who they believed the bad guys were in the fight. By reasonable deduction, the guys Google were saying were the bad guys must mean they were really the good guys.

With all its billions of dollars, there was no outlet Google hated more than tiny, but courageous *Breitbart News*.

I was going to contact *Breitbart* and turn over all the information I could get my hands on.

I'd been troubled by the "Covfefe" deception. But that incident had seemed foolish by comparison, almost Keystone Cops level shenanigans. Now I was looking at a cover-up of the deadliest mass shooting in modern American history, as well as obfuscation of a civil war in the Middle East.

I had become an enemy of Google.

CHAPTER EIGHT

The Blacklist

It was the fall of 2017 when I was sitting at my desk.

At the time Google was publicly telling the world that it didn't interfere with their "organic search" results. But as a normal full-time engineer with access to Google's corporate Internet I was able to simply type in blacklists. The first thing that popped up was something called the YouTube controversial blacklist

This blacklist contained search terms that were banned for videos on YouTube. What was strange about this blacklist was that a full 50% of the blacklist was dedicated to the Las Vegas massacre.

This is how it started out:

```
# YoutubeControversialTwiddler Query Blacklist.
# The queries in this file will force the
controversial twiddler to be
# triggered.
# This file will be parsed by
superroot::StringLookupModule. The data format is #
plain text. Each line contains one query, without
leading/trailing/duplicated # whitespaces. Each term
is delimited by a single space.
  64 old stephen paddock
  64 stephen paddock
  64 steven paddock
  64 years old stephen paddock
```

```
64 yr old stephen paddock
about stephen paddock
adress trump las vegas
anti trump paddock
anti trump stephen paddock
anti trump steve paddock
anti trump vegas
anti-trump paddock
atack las vegas oktober 2017 hoax
atack lasvegas
ataq las vegas
ataque lavegass
atephen paddock
attack a las vegas
attack from the left las vegas
attack in las vegas
attack in las vegas gun man
attack in las vegas live news
attack in the vegas
attack las veags
attack las vegas trump
```

And as I scrolled past pages and pages of banned search terms related to the Las Vegas shooting massacre I started noticing other mass casualty events. Words like "NYC shooting today," "Manhattan attack," "lower Manhattan attack," "suspect fleeing NYC attack," "suspect flying NYC shooting," "suspect driving NYC truck," etc. etc. etc.

```
truck
truck attack
truck incident
terrorist attack
terrorist
ramming
truck ramming
nyc attack
nyc shooting
nyc shooting today
nyc truck
```

```
nyc truck attack
nyc truck incident
nyc terrorist attack
nyc terrorist
nyc ramming
nyc truck ramming
manhattan attack
manhattan shooting
manhattan shooting today
manhattan truck
manhattan truck attack
manhattan truck incident
manhattan terrorist
manhattan ramming
manhattan truck ramming
lower manhattan attack
lower manhattan shooting
lower manhattan shooting today
lower manhattan truck
lower manhattan truck attack
lower manhattan truck incident
lower manhattan terrorist attack
lower manhattan terrorist
lower manhattan ramming
lower manhattan truck ramming
new york attack
new york shooting
new york shooting today
new york truck
new york truck attack
new york truck incident
new york terrorist attack
new york terrorist
new york ramming
new york truck ramming
```

And then further on in the list are things such as "live news about the shooting in Texas," "live news on shooting in Texas," "name of shooter," "name of the Texas shooter," "Sam Hyde shooter," "shooter Devon Kelly," and on and on.

I would say as a rough estimate that a full 95% of all of the blacklist terms involves information around mass shootings and what's even stranger is that a large percentage of the banned search terms includes things like "Crisis actor caught," "crisis actor David Hogg," "crisis actors Las Vegas" "how to spot a crisis actor," "Sandy Hook crisis actors compilation."

florida shooting conspiracy
florida shooting crisis actors
florida conspiracy
florida false flag shooting
florida false flag
fake florida school shooting
david hogg hoax
david hogg fake
david hogg crisis actor
david hogg forgets lines
david hogg forgets his lines
david hogg cant remember his lines
david hogg actor
david hogg cant remember
david hogg conspiracy
david hogg exposed
david hogg lines
david hogg rehearsing
florida shooting conspiracy
florida shooting crisis actors
florida shooting crisis actor
florida shooting false
florida nightclub shooting false flag
parkland shooting hoax
parkland school shooting false flag
parkland crisis actors
parkland false flag
parkland actors
parkland conspiracy
parkland hoax
parkland shooting false flag
parkland school shooting conspiracy
florida shooting actors

```
florida school shooting conspiracy
florida school shooting crisis actors
florida school shooting fake
david hogg
parkland
david hogg california
david hogg cant remember his lines
david hogg crisis actor
david hogg video
crisis actors
crisis actor david hogg
hogg
david hogg video california
false flag
crisis actor
crisis actors david hogg
```

The list of blacklisted terms just kept going. If I hadn't read it with my own eyes I wouldn't have believed it. This was censorship on an unbelievable level. With all the information coming at everybody so fast, would people really notice if some information just "vanished"?

```
gop train crash
gop train crash assassination
gop train crash assassination attempt
gop train crash attempted assassination
gop train crash attempted killing
gop train crash theory
attempted assassination of gop congress members
intentional gop train crash
gop train
train crash
911
```

And there was more to come, suggesting there was truth behind the claims of many groups that their information was being systematically suppressed.

```
--> cancer cure
--> cure cancer
```

```
new york
new york bombing
port authority
new york explosion
new york city
nyc
new york bomb
port authority bus terminal
bomb
```

What. the. Fuck. Yes. Google blacklisted "cancer cure." And you should also know that they blacklisted "cure cancer," just in case you tried it other way. Want to find information about Black Lives Matter on YouTube?

Sorry, that term has been blacklisted.

How about something as sacrosanct as a country's constitution?

```
The 8th Amendment of the Constitution of Ireland
```

Nope—Google banned that too.

Think about the ramifications of that, Google is a huge global company and is putting in blacklisted terms so that people can't search for their country's own constitutional amendments!

It was at that point I flashed back to my Russian girlfriend Marianna in 2011. She had come from the failed Soviet Union as an immigrant to the United States. She was arguing with me:

> **Marianna:** "Zach Google will be the MOST evil company in the world."
> **Me:** "No it's not. Google represents exactly what is right in the world . . . "

Marianna called it exactly! How naive I was. If Marianna ever reads this: Hey you were right, kiddo. Thanks for all the long discussions on the Soviet Union back in the day.

I continued scrolling down the blacklist. I found celebrity deaths like "Anthony Bourdain death" and "Kate spade suicide":

```
Anthony Bourdain
Anthony Bourdain death
Anthony Bourdain suicide
Anthony Bourdain murder
Anthony Bourdain pizzagate
Anthony Bourdain pizza
```

```
Tony Bourdain
Tony Bourdain death
Tony Bourdain suicide
Tony Bourdain murder
Tony Bourdain pizzagate
Tony Bourdain pizza
Kate Spade
Kate Spade death
Kate Spade suicide
Kate Spade murder
Kate Spade pizzagate
Kate Spade pizza
Kate Spade Bourdain
Spade Bourdain
Bourdain Spade
Pizzagate anthony bourdain
Pizzagate tony bourdain
Pizzagate kate spade
```

What exactly was going on here?!?!

What else could I find? I decided to search more and look for other blacklists. The next blacklist I found was called, google_now_black_list.txt which included sites like the mainstream thegatewaypundit.com:

```
thelibertarianrepublic.com/
usasupreme.com/
dailyheadlines.net/
investmentwatchblog.com/
--> thegatewaypundit.com/ <--
madworldnews.com/
vdare.com/
conservativetribune.com/
```

Then later I saw they blacklisted glennbeck.com, truepundit, louderwithcrowder:

```
usanewsflash.com/
webdaily.com/
msfanpage.link/
truthandaction.org/
```

```
amaziness.net/
--> glennbeck.com/ <--
thenewsclub.info/
news.grabien.com/
ilovemyfreedom.org/
--> louderwithcrowder.com/ <--
--> truepundit.com/ <--
rightwingwatch.org/
milo.yiannopoulos.net/
--> mediamatters.org/ <--
conservativefighters.com/
godlikeproductions.com/
```

This "Google Now" blacklist was to protect users on Google's news platform on their Android Phones. As people adopted new phones, they were memory-holing information.

For example, let's say you had an android phone and you were using the phone's news widget to access your news. Well then there was a whole class of journalists that Google had secretly determined should not appear on your news stream. Is that fair?

Some weeks later I would discover my third blacklist.

The way I found this list was by answering a call for help from a site called e-catworld.com. They wanted to get uncensored. For some reason, Google had put them on a secret list and therefore E-catworld.com was not able to rank in the search results at all.

I found this cry for help quite haphazardly.

I simply happened to go to this website to look at what they post. I do this occasionally for this website. But when I went to them in November of 2017, I wasn't able to find the website at all. And I couldn't remember the name e-catworld.com. After trying a bunch of search terms I stumbled upon e-catworld.ORG, which had a huge banner on the landing page saying that Google had delisted the site completely.

I was able to confirm this. Not even typing in the exact site "e-catworld.com" could bring it up.

What was e-catworld? This website was a news site about low energy nuclear reactions, a.k.a. "cold fusion" research. Google had decided that for some reason this news website was now banned from all its search results.

Since I had access to Google's entire search corpus, I simply did a search for this website "E-catworld.com." A link to a blacklist was the only result. This blacklist was called page_level_domain_blacklist.pdf Here's what it looked like:

```
Reported Issue, Assigned to pq-contmon@google.com
Please review go/pqescalate and enter the following
information:

1) Is this a page or site-level request?
Site
2) What is the site? http://www.cnn.com/2017/10/02/
us/las-vegas-attack-deadliest-us-mass-shooting-
trnd/index.html http://abcnews.go.com/US/wireStory/
las-vegas-attack-deadliest-shooting-modern-
us-history-50227779 http://www.foxnews.com/
politics/2017/10/02/las-vegas-shooting-lawmakers-
condemn-senseless-attackthank-police.html http://
www.express.co.uk/news/world/861237/las-vegas-
shooting-mandalay-bay-attack-emergencyphone-
number-hotline-terror-latest http://www.express.
co.uk/news/world/861210/Las-Vegas-shooting-what-
happened-Mandalay-Bay-terrorlatest-attack-news-
updates http://www.express.co.uk/news/world/861328/
Las-Vegas-shooting-updates-latest-news-shooter-
deathtoll-victims-Mandalay-Bay-attack http://www.
independent.co.uk/news/world/americas/las-vegas-
shooting-woman-told-crowd-youre-allgoing-to-die-
before-attack-brianna-hendricks-marilou-a7978821.
html http://www.reuters.com/article/us-usa-lasvegas-
shooting/gunman-kills-at-least-50-wounds-200-in-
lasvegas-concert-attack-idUSKCN1C70FU http://www.
bbc.co.uk/news/av/world-us-canada-41471532/las-
vegas-shooting-witnesses-describe-attack http://
heavy.com/news/2017/10/mandalay-bay-las-vegas-
boulevard-active-shooter-shooting-victimsattack/
http://www.bbc.com/news/av/world-us-canada-41471532/
las-vegas-shooting-witnesses-describe-attack
```

And then toward the bottom of the document, there was an entry for e-catworld.

```
http://world.breaking.e-catworld.com/news/las-vegas-
  shooting-50-people-killed-in-mandalay-bay-attack
```

Yet . . . this was impossible. As many can recognize, this is a bad web address that belongs to no one, especially not e-catworld.com, which has never covered general news as indicated in the url nor do they have the subdomain "world.breaking."

So what's with the weird censor list from Google above?

As you can see in this list there are lots of news pages covering the Las Vegas shooting attack.

Looking at this list, it's clear that EVERY MAJOR news organization got at least one of their articles about the Las Vegas massacre blacklisted from Google search.

For whatever reason, Google absolutely did not want you to find stories about the Las Vegas shooting.

It seemed obvious to me that since e-catworld.com was not a general news site that their appearance on a blacklist was sabotage. Either insiders doing it (I later found out that Google had a secret cold fusion lab) or someone on the outside who knew of an exploit on Google's systems and was infiltrating the data via open sources of truth.

I was not going to let Google censor science about free energy systems. I replied to the bug report and asked:

```
"Hi there, is this list above the reason that
e-catworld.com just got de-ranked from Google's
engine?"
```

The response was:

```
"No . . . this is completely unrelated as far as I
am aware"
```

Another engineer chimed in with a response:

```
"Rob is correct, changes to ad serving and changes
to search ranking are done completely independently,
and by different teams.
```

The bug report I had re-opened was then closed.

Frustrated, I created a new bug report against the Google search engine asking why this website, e-catworld, had been taken down and whether this had the possibility of corporate espionage. I actually wrote "espionage" in my email in order to get the attention of the lawyers in the company. It worked because I got a letter from an attorney that next day saying that I shouldn't be using such terms in company communication. I immediately apologized and thanked him for bringing it to my attention. And then, without further action on my part, the website e-catworld.com was restored back to the Google search index.

A small but important victory for this science tribe.

Having found three blacklists, I decided that such evidence must be preserved for at least my records to verify that all this was really happening. But this was raising a serious question in my mind which needed answering. Was Google becoming evil? Instead of organizing the world's information and making it universally accessible, Google was installing itself as the secret gatekeeper to your access to information.

I hit the "Print → Save as PDF" button on my Chrome browser and downloaded them for safe keeping.

CHAPTER NINE

Approaching *Breitbart*

I started calling people, journalist friends of mine, individuals I'd come to know over many years with a different take on things, who might have an idea on how to contact somebody at *Breitbart.*

I know the conservative media likes to paint Silicon Valley as a group of people who uniformly believe the liberal line. But it goes against the non-conformist tendencies of your typical engineer. I agree with the belief the majority of those working in Silicon Valley are of the liberal mindset. However, there's a significant minority of us who are not.

Think about it.

We're the people who want to create new products and industries. Don't we reject the status quo as a matter of principle? When we hear authority talking, are we saying to ourselves, "Our wise and beneficent leaders have finally achieved perfection!"? Does the creative person look at the world and say, "Hey, everything is just fine." NO. In fact, we think "Why are they being so stupid? I could do better in my sleep!"

Some of my contacts on Twitter were able to introduce me to Matt Tyrmand, who works at *Breitbart* and sits on the board of Project Veritas. I explained who I was, and he said, "This seems too big for us. You should probably contact Project Veritas."

I knew Project Veritas, the group founded by James O'Keefe, which practiced undercover investigative journalism, usually resulting in some exposé they'd post on their website and on YouTube. O'Keefe had first attained fame by going after a group in 2009 called ACORN (Association of Community Organizers for Reform Now), a group which Barack

Obama had supported and given several government contracts to help with the 2010 census. According to O'Keefe, at its height, ACORN had more than five hundred thousand members and more than 1,200 chapters in more than eleven American cities.[1] O'Keefe had gone undercover himself, posing as a pimp, along with a woman posing as one of his prostitutes. The pair showed up at the Baltimore ACORN office and claimed they wanted tax advice on how to open a brothel. As O'Keefe recounted on his website:

> James O'Keefe, founder and President of Project Veritas, along with Hannah Giles, posed as a "pimp and prostitute," and visited an ACORN office in Baltimore, Maryland, seeking advice on how to obtain a loan to start a brothel. Astonishingly, this unusual and illegal request was not met with even the slightest of hesitation, as the ACORN tax expert was all too eager to give the professed "pimp and prostitute" the advice they sought. In fact, the ACORN employees in the Baltimore office went on to give advice on how to avoid the police and how not to get into trouble with an abusive pimp.[2]

James is a tall, clean-cut, good-looking young guy with a cherubic face. His disguise was a dark blue suit, light blue shirt with white collar, red tie, fur coat, pimp sunglasses, and a black hat. Hannah is an attractive brunette, but the big blue loop earrings, tight brown leather pants and skimpy leather top didn't look convincing to me at all. To my eye, the two of them looked like a couple of college kids dressed up for a Halloween frat party.

But in addition to the requested tax help, they received advice on avoiding the police and abusive pimps. However, the helpful Baltimore ACORN staff wasn't done aiding the eager young pimp and his girl.

> Going from the ridiculous to the sublime, the ACORN employees freely gave advice on how to claim as dependents 13 "very young" girls, purportedly trafficked from El Salvador to be prostitutes. Indeed, she went on to warn O'Keefe and Giles that: "with your girls [the aforementioned 13 underage Salvadoran prostitutes], you tell them to be careful. Train them to keep their mouths shut. Always keep your eyes in the back of your head."[3]

As a result of the investigation by Project Veritas, the IRS and the Census Bureau severed their ties with ACORN.[4] In the ensuing years, Project Veritas would go on to do in-depth undercover investigations of the misuse of government stimulus money for Green Jobs, Obama phones, voter fraud, the Veteran's Administration, teacher unions, and border security.[5] For these stories, which in previous decades would have been lauded in the press, Project Veritas was often disparaged in the mainstream media.

Breitbart gave me the contact information for a Project Veritas operative, I contacted him, and he made plans to come to San Francisco. In order to get the story right, I thought it was necessary for the operative, Michael Combest, to spend many hours with me as I spelled out the story. I couldn't give to him the entire story in an hour or two interview.

* * *

Mike was a great guy with whom to share these stories. He was super interesting to talk with and funny.

However, the whole situation was immensely stressful, meeting with this operative, disclosing information about what I believed to be the crimes of my employer, and then getting up the next day to go work at Google. As I'd take the Google bus to work and watch the Northern California scenery go by, I'd think about how I was figuratively sticking a knife in the back of one of the wealthiest companies on the planet. I was terrified about all the ways it might go wrong.

I'd walk into work, pass the cubicles of my fellow employees, and think, *Wow, I'm involved in this operation that's likely to affect the business of every one of my co-workers. If they ever discover what I've done, they're going to hate me for it.* I felt so much anxiety that I didn't want to hang out with any of my co-workers for fear I'd give myself away.

Instead of engaging in community activities with the team, I began to pull away. It seemed to me they were indoctrinated, suffering from massive white guilt, and pathologically altruistic. There was a kind of certainty in their opinions, a fear of Trump voters, as if they were subhuman, and the very act of speaking with one risked some sort of contamination.

And yet my nights with Mike, where I unburdened myself of these feelings, were glorious, as if the simple act of verbalizing the truth was putting something good into the world.

By talking with Mike, I was able to see the outline of Google's misguided plans in a clearer light. When you put it all together, it seemed to be six simple, interconnected programs. I was able to work on what's generally called an "elevator pitch," meaning you could give a good overview to a person in a few minutes.

At the top of the pyramid was "Machine Learning Fairness," the algorithmic tool that would be rolled out to Google searches, Google News, and YouTube. This allowed the executives to monitor whether the information being put out was consistent with the ideology Google wanted to promote.

From that point you needed what came to be known as Google's EAT score, which stands for Expertise, Authoritativeness, and Trustworthiness.[7] EAT was introduced in 2018 and used to down-rank a number of so-called "alternative health news," such as suppressed cancer cures or parents expressing concerns about vaccine safety. One of their standards for accuracy was having a Wikipedia page,[8] which even middle school librarians generally tell their students to avoid. Google's EAT scores were an exceptionally flexible program, giving Google the ability to immediately lower the visibility of information which ran counter to their preferred narrative. I'd argue that the EAT scores are one of the most powerful tools Google has for censorship, as one can simply claim a source does not have "expertise," "authoritativeness," or "trustworthiness." Who could challenge these decisions if they take place inside a machine, free of apparent human influence? Even if a decision by the algorithm is shown to be wrong, they can simply say, "We got bad information, but we tried to do the best we could. Sorry!"

Another tool was the "Trend Identification of Social Media," which would provide a real-time boost to certain social posts in accord with Google's beliefs.

An additional tool was called "Super Root," and it was a re-ranking system using an application called "Bit Twiddler."

Finally, after the system was complete, Google would use its "Purple Rain" application to monitor developing news stories and de-boost news, which was not in accord with their narrative. I believe the system was first deployed to squelch alternative narratives of the Las Vegas Massacre.

With these six programs, Google would have full dominance on the digital information battlefield.

When you start to understand the architecture of information control that Google was employing to control the American electoral system, it can take you to some very dark places. I began looking at things like the Charlottesville Riot, the Trayvon Martin case, and the Sandy Hook shooting in a different light. I was beginning to seriously wonder, *what was real*?

I think it's important to stop for a moment and discuss a concept I believe has been purposely misrepresented, the idea of a "false flag." It doesn't mean the event never happened. What it means is that some facts are being misrepresented in order to create a different narrative. Although

I can't prove it, I believe it's being done intentionally. Whether it's to generate more clicks for news outlets, or to further a political agenda, I don't know.

Let's take the Las Vegas massacre as an example. There are some simple questions to be asked, such as whether members of the Saudi royal family were occupying the Four Seasons hotel, and if so, was there any connection to what happened? Instead of questions along those lines, or delving into subsequent events in Saudi Arabia, we got a debate about "bump stocks" on guns. Were we being diverted like in a magician's trick to not look at the other hand?

I can't tell you the answers. But I know they're important questions.

The people I was working with weren't interested in asking such questions. I found it difficult to have simple conversations with them. They'd totally bought into the system and believe they're on the right side of history. They genuinely think the same country that elected Barack Obama twice, is now filled with neo-Nazis, who are trying to use "fake news" to take over the government. I couldn't stand their anti-Trumpism. I wanted to yell at them that they were ruining the country with their paranoia.

But I avoided betraying my real thoughts.

Because even though I knew they were smart, I realized they'd been brainwashed. I'm just not built that way. I'm an anti-collectivist to the very center of my being. If somebody gets too comfortable in what they believe, I'll immediately move to the other side and see if I can find the flaws. The more collectivist the thinking—the crazier corporate groupthink that comes down the pike—the more I pull away.

I was working on non-political projects for Google at the time, creating applications which would allow YouTube to be on your television, and then using the same for the Nintendo Switch, Xbox One, and PlayStation 4. I can always lose myself in an engineering problem, and it's common for those working on the projects to become consumed with them and be anti-social. That was the game I tried to play during those months.

But I knew I was getting radicalized by these discoveries. The MOMA system, allowing any Google employee to see what anybody else in the company is doing was still open, and I was still finding documents, making PDFs of them, but then worrying that if I downloaded them there would be a digital trail. I'd take pictures of the documents with my cell phone, then try to hand them over to Mike, but the quality was just terrible. Finally, I decided to simply print out the documents, knowing I was leaving evidence behind.

I told Mike, "If you trust James O'Keefe, then I'm going to trust him. I'm going to believe I can hand you these PDFs and you're not going to leak them out in a way that can be traced back to me. Because I'm very concerned about my safety if it becomes known that I did this. I just want the truth to get out. Because I know good people will do the right thing with this information."

Mike was happy with my decision, took all the documents, and said he couldn't wait to get back to the Project Veritas office and share the information directly with James O'Keefe. I breathed a sigh of relief. I'd done what I could as a good and loyal citizen of my country.

But what I didn't know at the time was that my story didn't yet fit the Project Veritas model.

It was as if I'd laid bare my soul and the universe in it's majestic indifference, just shrugged.

* * *

Then out of the blue, Mike called. But it wasn't for Google. Mike asked if I wanted to help him in a sting operation against Twitter. It sounded fun, but I was too stressed out to consider it. Michael's work landed an important story, getting a Twitter engineer, Pranay Singh, to talk about banning conservatives through the use of an algorithm.

> "Just go to a random [Trump] tweet and just look at the followers. They'll all be like guns, God, 'Merica, and with the American flag and the cross," declared Singh, who was secretly recorded by Project Veritas reporters. "Like, who says that? Who talks like that? It's for sure a bot."
>
> After being asked whether he could get rid of the accounts, he replied, "Yeah. You just delete them. But, like the problem is there are hundreds of thousands of them. So, you've got to like, write algorithms that do it for you."[9]

I was pleased Michael was able to land such a big story, but it seemed like small potatoes to me. Although it made some sense to go after Twitter, I thought it was missing Google at the center of the spider's web of censorship.

* * *

What do you do when you commit yourself to a course of action, and then nothing happens?

I had damning information about Google, and I'd offered it to the only two entities I trusted, *Breitbart* and Project Veritas. I couldn't understand why Project Veritas wasn't interested in my story. Michael tried explaining to me the way Project Veritas worked, more of a *60 Minutes* style with packaged fifteen minute stories with a nice clean narrative. By comparison, I had a sprawling story of critical race theory being infused across multiple platforms, algorithms, and operating systems.

Nobody knew what I'd done and it didn't seem like I'd be causing trouble for anybody, saving Google and every employee in their company. You might call me an unindicted traitor to my company, and that's just how I felt.

CHAPTER TEN

The YouTube Shooter

I went back to work among my Trump-hating co-workers, took on some interesting projects, and just kept collecting my ridiculously large paychecks. My own investigations continued, trying to understand what was really going on in the world, but resigned to keep my research hidden from everyone else at all costs.

That was until the early afternoon of April 3, 2018.

I worked at Building 901 on Cherry Avenue in San Bruno, California, the YouTube headquarters. The architecture was new age, sleek and modern, with a gigantic front that opened up like a fish mouth that looks like it's going to swallow you. The tops of the building resembled rolling hills, with grass on the top, so if you were looking at it from a satellite it would just seem to be part of the landscape. The grounds were two blocks long by a block wide, with several buildings, a large lawn, and wooded area with trails for employees to take their dogs on walks during lunch.

Yes, it was a California crunchy-granola socialist paradise.

But I wasn't safe from the outside world.

I was about to be in the middle of a god-damned workplace shooting.

Like a good little corporate employee, when I heard the fire alarm go off, I thought, *Could this be a false flag?* But I quickly discounted the thought as being paranoid.

I picked up my electric skateboard and headed with the rest of the employees toward the nearest exit, which was at the back, and exited to a large wooded back area.

I flopped my skateboard down on the ground and began skating down the hill, ahead of the massive wave of footwalkers behind me. As I did so, I noticed that there were three people standing at the bottom of the path, which dumped into the patio courtyard. The same courtyard where I had lunch nearly every day.

There was someone with a five o'clock shadow lying on their back. I thought to myself that it was odd that someone was lying down on the cement. There was broken glass from the doors leading into the building. Looking closer at the person laying on the ground I noticed there was a red stain on their stomach.

A man nearby yelled incoherently, pacing back and forth. He had a beer gut, was fat and bald, and had old, dirty clothes on. He yelled out, "Do you want to shoot me, too? Do you want to shoot me, too?"

It was a surreal moment, which got even stranger when the patio door leading to the outside swung open and an armed policeman with what looked like an AR-15 assault rifle burst through.

At this point I realized this was not someplace I wanted to be. "Shit! Shit! Shit!" I thought. "This is the real deal! This is an active shooter situation!"

I threw my electric skateboard on the cement walking path which led behind the nearby parking garage, and jumped on it, hitting the electric throttle to full and speeding away from the scene. My heart started pounding and I worried about being hit by a stray bullet.

As I sped away, I noticed that the wave of footwalking employees descending the hill were about to come face to face with armed police and an active shooting zone.

I yelled up to the human wave: "ACTIVE SHOOTER! ACTIVE SHOOTER!"

Some of the people understood what I was saying and started running away, hitting the fence around the area and climbing over it. After the past few months of me researching the Covfefe incident, Las Vegas, and the Saudi anti-corruption purge, among others, I thought to myself I should get some video of what was happening for use as evidence later of what had really happened. I rode on my skateboard to the parking structure, hoping to get to one of the top floors for a better vantage point. But there was a security guard and he told me I had to leave.

On the public street, I started live-streaming what was going on to Facebook as the police, ambulances, and fire trucks arrived. News reporters were gathering near the masses of people streaming out from YouTube, but few people were stopping to talk to the media. While I suspected that part of that was due to the fact most people are shy of talking in public, Google also had a strict policy about talking on camera and required employees to have a chat with the Public Relations department before granting approval. However, an active shooter situation is definitely out of the ordinary, but employees still seemed to avoid talking to the media after this event. A group of media reporters were stationed at the intersection of Traeger Avenue and Bayhill Drive, trying to get comments from employees as they walked past. NONE of them I saw were talking to reporters. I was likely the two hundredth person that the media people asked, no joke. As I was walking by, I made eye contact with one of the reporters and he took that as a sign of encouragement. I started talking to him.

Since I'd been on *America's Greatest Makers*, and done video for my Kickstarter campaign, I was comfortable in front of the camera. The media swarmed me and immediately I became the go-to person on the YouTube shooter.

Suddenly my phone was ringing with different news outlets like the *New York Times* calling me for comment and I was trying to keep up with it all.

But then there started to be other media reports that the shooter was a woman, and people started wondering if there were two shooters active at YouTube.

The *New York Times* called me back and the reporter started ripping into me. "You're a liar! Who are you? Are you a disinformation agent?"

Yes, a reporter for the *New York Times* asked if I was a disinformation agent. Had the mainstream media finally reached peak insanity? "What are you talking about?" I asked.

"You said the only victim was a man. But the only victim was the shooter. And the shooter was a woman. So why did you tell us the shooter was a man?"

"Look," I said. "I'm sorry, but I saw a man. And I reported it as a man on my Livestream and I work for YouTube. I'm not an intelligence operative."

We hung up, but the reporter called back a while later. "We're not exactly sure what's going on," he said. "But what you told us about the shooter coming in through the parking garage checked out. So, we're really confused on this story, now."

I was thinking to myself, *who the hell is the shooter*? Later I did some research and it got even weirder. This is from the Associated Press story on the incident:

> The woman suspected in the shooting at YouTube headquarters Tuesday was a 39-year-old San Diego resident.
>
> Two law enforcement officials identified the suspect as Nasim Aghdam.
>
> Aghdam was quoted in a 2009 story in the San Diego Union-Tribune about a protest by People for the Ethical Treatment of Animals against the use of pigs in military trauma training. She dressed in a wig and jeans with drops of painted "blood" on them, holding a plastic sword at the demonstration outside the Camp Pendleton Marine Corps base.[2]

Have you been listening to me enough to pick out what's unusual about this account? Nasim's video clearly showed what looked like a man.

Let's start with the name, Nasim. I understand most of my readers won't be Muslim, so the vast majority can be forgiven for not knowing whether Nasim is a male or female name. According to the Muslim website Urdu Point, Nasim is a male name which means "fresh air" in Arabic.[3] In fairness, there's some indication Nasim can be used as a female name, but it is more commonly used as a male name.

Was the rampage predictable? According to Nasim's father, it was, and he warned the police that his son/daughter might be planning to attack YouTube.

> The father of a woman suspected of shooting three people at YouTube's headquarters says she was angry at the company because it stopped paying her for videos she posted on the platform.

> Ismail Aghdam told the Bay Area News Group that he warned police his daughter, Nasim Aghdam, might go to YouTube because she "hated" the company.
>
> Ismail Aghdam said he reported his daughter missing on Monday after she did not answer her phone for two days.[4]

Nasim's father alerted the police to the potential for violence by his child. But when I did some research on Ismail Aghdam, I came up with some intriguing claims. Some posts suggested he'd worked for Iranian intelligence before coming to America. I couldn't verify those accusations, but I simply put it in my mental file. An article from the *Seattle Times* painted this picture of the shooter:

> Aghdam was prolific at producing videos and posting them online. She exercised, promoted animal rights and explained the vegan diet, often in bizarre productions with elaborate costumes or carrying a rabbit.
>
> She posted the videos under the online name Nasime Sabz, and a website in that name decried YouTube's policies, saying the company was trying to "suppress" content creators.[5]

Let's take a closer look at Nasim, starting with the name the shooter used online. Was Nasime an attempt to feminize Nasim? It's difficult to know because when one searches the name and its origin, nothing comes up. In other words, the name Nasime seems to be a unique creation of the shooter.

Even if we can't come to a conclusion about the gender of Nasim's name, we've got a vegan animal rights activist with a passion for working out, who goes and shoots up a tech company.

Not really the liberal narrative of an angry white male shooter that the mainstream media would prefer.

The police response was also similarly confusing. Was it simply bad policing, something nefarious, or the police not wanting to be accused of racial profiling? The father claims he reported his "daughter" missing to the police and that she might be headed to YouTube. The Associated Press reported the father's claim that:

> He said the family received a call from Mountain View police around 2 a.m. Tuesday telling they found Nasim sleeping in a car and he warned them she might go to YouTube.

> Mountain View Police spokeswoman Katie Nelson confirmed officers located a woman by the same name asleep in a vehicle in a Mountain View parking lot Tuesday morning.
>
> She says the woman declined to answer further questions. Nelson did not respond to a question about whether police were warned Aghdam might go to YouTube.[6]

Could it simply be that in the wake of an unthinkable tragedy that all the warning signs appear to be so clear? Am I just Monday morning quarter-backing the police response? It's possible.

Probably the best article I later read on the YouTube shooting and the media reaction was put together by Rachel Blevins at the Free Thought Project. It listed five reasons the YouTube shooting disappeared from discussion in the mainstream media.[7]

The first reason was: "As a woman, a peaceful vegan, and a PETA advocate, the suspected shooter is the opposite of the typical 'mass shooter' profile promoted by the mainstream media."[8]

The second reason was: "The shooter reportedly used a handgun, which also deviates from the mainstream narrative that all mass shooters used high-powered rifles."[9]

The third reason was: "The shooting happened in California, a state that has already enacted some of the strictest gun control laws in the nation."[10]

The fourth reason was: "The police were warned about the shooting by the suspect's father in advance, and they did nothing, something the media tends to cover up."[11]

The last reason was: "The shooter blamed YouTube for censoring and demonetizing her videos, a problem alternative content creators experience on a daily basis that mainstream media tries to pretend doesn't exist."[12]

I agree with all the reasons listed above, although my complaint is much simpler. I stood over the body of a dead man. But the media told me I was standing over a dead woman.

I felt as if I was living in George Orwell's book, *1984*, in which the three slogans of the Party were: "War is peace, slavery is freedom, and ignorance is strength."

I couldn't help but wonder if Orwell could have gotten away with a fourth slogan for Big Brother, "a man is a woman." Please ignore the male physique, large Adam's apple, and several days' growth of beard on her face.

CHAPTER ELEVEN

Project Veritas Returns

April, May, and June of 2019 were months filled with repeated waves of censorship from YouTube.

Huge swaths of conservative content were getting "demonetized," destroying the livelihoods of content creators as their revenue streams on YouTube dried up. Because of the silence from Project Veritas, I assumed they didn't want to follow up on the "Machine Learning Fairness" story I'd presented to them. I couldn't understand this because they were breaking other stories about Big Tech, like the social media company Pintrest flagging Bible verses as pornography.

By the time June 2019 came, I was thoroughly demoralized by this state of affairs, and how everything I'd put at risk for Project Veritas seemed to be for nothing. I thought there would be no disclosure.

The problem was that I had to wake up every morning and head into work for a company I felt was the enemy of America and her people. I could not bear to be a part of this evil system of Google and YouTube, even though I was making $250,000 a year with them.

It was in the first week of June 2019 that I realized I was finished with the company. During a weekly meeting with my manager he was blabbering on about the new "roadmap," and I knew in my heart I wouldn't be around for it. I simply couldn't carry the burden anymore.

I interrupted him. "Hey, Henry, I'm sorry. But I just can't do this anymore. The YouTube censorship is just too bad and is against everything I believe. I'm going to exit the company. I'll give my official two week notice next week. That should give you some extra time to find my replacement."

And with that, I was quitting Google.

A week later, I was in my apartment, writing my resignation letter, when I got a call from Joe Halderman, one of the executive producers at Project Veritas. "Hey Zach," he began, "We have this transcript and we want to get your opinion on it. Can we send it to you and have you keep it secret?"

"Sure," I replied, not believing the synchronicity of the moment. Within ten minutes I'd received the transcript and was reading through it. My jaw dropped. It was a transcript of one of the Google executives, Jen Gennai, about whom I'd sent information to Project Veritas. She was quoted as saying:

> We all got screwed over in 2016. Again it wasn't just us. It was, the people got screwed over, the news media got screwed over, like everybody got screwed over. So we've rapidly been . . . how do we prevent this from happening again?
>
> We're also training our algorithms. Like, if 2016 happened again, would we have, would the outcome have been different?
>
> Elizabeth Warren is saying we should break up Google. And like, I love her, but she's very misguided. Like, that will not make it better, it will make it worse. Because all these smaller companies who don't have the same resources that we do will be charged with preventing the next Trump situation. It's like, a small company cannot do that.

I was stunned. Was this actually happening? Everything I had hoped for was suddenly materializing in front of my eyes. Project Veritas had given me the results of *their* sting operation.

A few minutes later, Joe Halderman texted me, "So, what do you think of the transcript?"

I called him back immediately and he had James O'Keefe join in on the call line. I started talking fast. "This is exactly what I was talking about. And you actually got a Google executive to admit everything?" Joe and James said yes. I continued on in my excitement, talking about my concern that Google and the mainstream media were working in lock-step to undermine our society and censor the competitors of the corporate media.

When I finished, James responded, "What you just said was amazing, Zach. Would you like to come here to New York and say it on camera?"

I was stunned by the suggestion. This was the opportunity of a lifetime to tell the American people why they were being censored, but more importantly, the nuts and bolts of how they were doing it.

I told James I just happened to be writing my resignation letter and he'd need to give me time to think about it. If I did do it, I'd probably want to be anonymous.

And with that, I got off the phone, finished my resignation letter, and headed into work.

* * *

This is the resignation letter I sent on Monday, June 19, 2019. (Since I sent the letter on a corporate server I no longer have access to it. This is my best recollection of what I sent to them, as I provided to my co-author in October 2020.)

To the Cobalt team here at YouTube:

> I'm excited to announce that I'm moving on from YouTube. The last three years have been an incredible time of growth and I have been fortunate to work with some of the smartest people I have ever known.
>
> During my time at YouTube I have had the fortune of bringing the YouTube app to three game consoles, the PlayStation 4, the Xbox, and the Nintendo Switch. These platforms (along with others from the Cobalt team) now constitute 10% of YouTube's entire traffic stream. What an honor to have that much impact at scale.
>
> Thank you all for a wonderful experience.
>
> Zach Vorhies.[3]

I was planning to sneak away from Google and YouTube, not letting them know I was providing their secrets to the public. I wanted to do my duty as an American citizen, but I didn't want to be reckless with my safety. One doesn't have to be a conspiracy theorist to know that if you publicize the secrets of a large corporation you put yourself at significant risk.

* * *

Project Veritas put together a twenty-five-minute video on Google, using the comments of Gaurav Gite, Jen Gennai, and me, the anonymous whistleblower, which they put out on June 24, 2019. The comments of Jen Gennai drew the most attention:

> The reason we launched our A.I. principles is because people were not putting that line in the sand. That they were not saying what's fair and what's equitable. So, we're like, well, we are a big company and we're going to say it . . . People who voted for the current president do not agree with our definition of fairness. We're also training algorithms, if 2016 happened again, would the outcome have been different . . .
>
> We got called in front of Congress multiple times. Like, they can pressure us, but we're not changing.[4]

When I first joined Google in 2008, I believed in their mission statement to "organize the world's information and make it universally accessible and useful." In fact, it's still on Google's website to explain what they do.[5]

But from Jen Gennai, the head of Responsible Innovation at Google, we heard a different explanation. They're going to tell the public "what's fair and what's equitable." She's also comfortable in noting that approximately half the country who voted Donald Trump into the presidency "do not agree with our definition of fairness." In the vocabulary of the playground, "it's my ball and if we don't play according to my rules, I'll just take my ball and go home." Seriously, do any of you have a couple billion dollars to build a search engine that's more in-line with Google's original and still current mission statement?

And finally, it's not just the current occupant of the White House with whom they have a disagreement, but our very system of government. How else can one interpret the cavalier way in which Gennai dismisses Congressional oversight?

"Like, they can pressure us, but we're not changing."

Is it because Google's lobbyists donate to both members of the aisle in Congress that they know the only thing that will be done is some phony hearings which will leave Big Tech free to do exactly as it pleases? Even liberal Senator Elizabeth Warren who ran for the Democratic nomination for president in 2020 doesn't escape the paternalistic superiority of Jen Gennai.

> Elizabeth Warren is saying that we should break up Google. And I love her, but she's very misguided. Like that will not make it better. It will make it worse. Because now all these smaller companies who don't have the same resources that we do will be charged with preventing the next Trump situation. It's like a small company cannot do that.[6]

Now, who would have ever thought that Senator Elizabeth Warren and President Donald Trump would be on the same side of an issue? But maybe it shouldn't be so shocking. Although Americans may disagree on their politics, both sides have traditionally had a reverence for our system which channels our passions onto a level playing field in which the best ideas of both sides compete and the people decide.

Google sought to fundamentally change that political dynamic. And they weren't even hiding it.

But if it wasn't clear from the statements of Jen Gennai or Gaurav Gite, I was going to make it crystal clear.

* * *

I was cloaked in darkness for the interview and my voice altered like a silver screen villain, but the words were mine. For the interview, I focused on Google's own internal documents, starting with their use of "Machine Learning Fairness" and how they defined "Algorithmic Unfairness."

> "Algorithmic unfairness" means unjust or prejudicial treatment of people that is related to sensitive characteristics such as race, income, sexual orientation of gender, through algorithmic systems or algorithmically aided decision-making.[7]

I showed their own documents which noted that something which is factually true, such as the predominance of men in CEO positions from a Google image search, could still be algorithmically unfair. Even if the results were not intended to be unfair, the fact that there was a discrepancy would meet the standard for algorithmic unfairness and be changed. How far would that silly standard go?

Machine Learning Fairness in action was even more ridiculous. In a Google search there's a function known as "auto-complete" in which you start typing an inquiry and the system makes several suggestions to help you before you've fully entered your inquiry.

For example, if you typed in the words, "women can" your top five suggestions would be "women can vote, "women can do it," "women can do anything," "women can be drafted," and "women can fly."[8]

If you typed in the words, "men can," your top five suggestions would be "men can have babies," "men can get pregnant," "men can have periods," men can have babies now," and "men can cook."[9]

This standard even applied in the political world. Typing in "Donald Trump emails," (even though Donald Trump has never used email), yields these results, "Donald Trump, Jr. emails twitter," "Donald Trump Jr. emails pdf," "victory Donald Trump emails," "Donald J Trump emails," and "Donald Trump campaign emails."[10]

And yet if you typed in "Hillary Clinton emails," which had been the subject of countless articles, it returned no suggestions.[11] (As of the time of writing, Google has changed this and now no results pop up for either.) As I said in my interview, "They're going to redefine reality, based on what they think is fair, and based upon what they want. And what is part of their agenda."[12]

Moving on from their algorithmic unfairness and machine learning fairness, Google would determine which news sources were credible, creating their own digital news ecosystem in which there would be a "single point of truth,"[13] across all their products, and conceivably across all major digital platforms.

With a situational awareness of which news sites they considered credible and which they did not, Google was in a position to down-rank or demote content from creators they considered dangerous to their narrative, like Dave Rubin or Tim Pool, former liberals who expressed at least some openness to discussing conservative ideas. As I explained to Project Veritas:

> What YouTube did is they changed the results of the recommendation engine. And so what the recommendation engine is, it tries to do, is it tries to say, 'well, if you like A, then you're probably going to like B.' So, content that is similar to Dave Rubin or Tim Pool, instead of listing Dave Rubin or Tim Pool as people that you might like, what they're trying to suggest different, different news outlets, for example, like CNN, or MSNBC, or these left leaning political outlets." [In other words, we'll give you what you don't like.][14]

When I explain this idea, many people struggle to understand it. Let me take it out of the political realm, as many may not know the ideological leanings of a Dave Rubin or Tim Pool. The idea of artificial intelligence the public was sold was that the system would look at you as an individual and give suggestions based on your searches or the materials you consumed. Let's imagine you liked to watch funny cat videos on YouTube. Your expectation is that the system will notice your viewing preferences and suggest more funny cat videos, or possibly other funny animal videos, maybe even babies doing funny things. That would be a natural progression, right? You get to be you, in charge of the technology, and the technology is only better serving your individual and legal preferences.

But let's say you finish watching that funny cat video and the next suggestion YouTube serves you up is a video on how cats can spread parasitic or bacterial infections to their owners. Or maybe the next set of videos involves dog owners explaining why dogs are better than cats. Or perhaps you get a string of grisly videos of some old lady who died with lots of cats, and by the time people discovered her body, the cats had started to eat her. Is the technology serving you, or are you being programmed to dislike cats?

Probably the most disturbing document I took from Google was one which boldly stated, "People (like us) are programmed."[15] It detailed the way we as regular consumers of information create our picture of the world, what we do wrong, and how they were going to reprogram us to be better human beings. Where have we heard similar claims throughout human history, and how many millions have died for our right to be imperfect human beings, you know, the kind without blonde hair and blue eyes? I'll let you answer that question, but here's Google's four-step description of the current process and their two additions to make you a better person.

The first step in this process was "Training data are collected and classified."[16]

The second step was "Algorithms are programmed."[17]

The third was, "Media are filtered, ranked, aggregated, or ***generated***."[18] (Bold and italics added by author.) Did you catch that? If nobody in the media has the correct view on a certain situation, Google reserves the right to "generate" such media.

The fourth was "People (like us) are programmed."[19]

The first of their two new additions was "Unconscious bias gets reinforced in the training data."[20]

The second addition to the process was "Unconscious bias affects the way we collect and classify data, design, and write code."[21]

Let's go back to our imaginary person who loves funny cat videos. Their "unconscious bias" in favor of cats will now be replaced with terrifying ideas of the diseases cats can spread to humans, why dogs are better than cats, and the image of an old lady being eaten by the cats she owned. Now, all those ideas may have their place in a discussion of the benefits of cats versus dogs. But why does one person, or one company, get to decide what you look at or think about? I made my feelings clear about what my former employer was doing:

> The reason why I came to Project Veritas is that you're the only one I trust to be able to be a real investigative journalist. Investigative journalist is a dead career option. But somehow, you've been able to make it work. And because

> of that, I came to Project Veritas, because I knew that this would be the only way that this story would be able to get out to the general public.
>
> I mean, this is a behemoth. This is a Goliath. I am but a David, trying to say the emperor has no clothes. And being a small little ant, I can be crushed. I am aware of that. But this is something that is bigger than me. This is something that needs to be said to the general public.[22]

Was I being crazy by going public? I was going up against one of the richest companies on the planet. But I was trying to be safe.

I was an "anonymous whistle-blower" and when I'd left Google, it was on good terms. I did not suggest when I resigned that I hated what they were doing as much as I did.

I was trying to follow the principle of Sun Tzu's *The Art of War*: "Let your plans be dark and impenetrable as night, and when you move, fall like a thunderbolt."[23]

However, I understood my plan could fail, and I might be revealed as the anonymous whistleblower. In fact, James O'Keefe even warned me I was putting myself in greater danger by NOT revealing myself as the whistleblower. "If you try to remain anonymous," he warned, "they'll have a target on your back."

I should have listened to James O'Keefe.

But I did make one very good move at the time.

* * *

One of the reporters I got to know at Project Veritas was Cassandra Spencer, a member of the investigative team who'd gotten the Jen Gennai interview.

Cassandra was a former public affairs officer for the US Army Reserve for seven years and had worked at Facebook. She'd started feeding information to Project Veritas in October 2017 and was fired in January 2018 without explanation, although it seemed to be based on some undercover reporting she'd done for a piece on Twitter in that same month.

Her involvement with Project Veritas marked a shift in the way they did things, moving from their previous strategy of sending people undercover in the target organization, often spending weeks or months to develop trust, to finding current members of the organization who had become dismayed by the company's practices. During her time with Facebook, she'd recorded many undercover videos and that story was to be published by Project Veritas in March 2019, a few months before my story went public.

Cassandra provided an interview to my co-author in November 2020, a day before the Presidential Election, and it was interesting to hear things from her perspective.

According to Cassandra, after the 2016 election, Project Veritas started to become a target because of their operations which targeted organizations with a liberal bent. Because of the high-profile of the stories they covered, a lot of their articles ended up in litigation. At the time of the interview, Project Veritas was involved in eight separate lawsuits.

As for the initial hesitation about the documents I'd presented to Project Veritas, Cassandra said, "If someone comes to you with information and it seems too on the nose, you think, is somebody trying to set me up?" She continued, "You have to be skeptical of information that's presented to you. Could Zach have manipulated those documents? None of us are document experts. And that's why we really wanted to get one of those executives at Google, like Jen Gennai, on video, confirming it."

Cassandra found the most damning statement from Jen Gennai was the cavalier way in which she dismissed suggestions from both liberals like Elizabeth Warren and conservatives like Donald Trump: that Google should be broken up. "And that's what's so aggravating," Cassandra told my coauthor. She continued:

> Nobody, the left, or the right, elected these Big Tech overlords. As much as I'm frustrated by our legislators not doing anything about these problems, I appreciated it recently when Ted Cruz went after Jack Dorsey of Twitter [In the weeks leading up to the 2020 election, Twitter had banned the *New York Post* over an article reporting on evidence from Hunter Biden's laptop suggesting Joe Biden had been involved in his son's foreign business dealings], asking, 'Who the hell elected you?' Because that's kind of the same reaction I had when Jen Gennai was saying these things. No one elected them.[24]

Of the impression she got from sitting for three hours in that noisy Mexican restaurant with Gennai, Cassandra said:

> The sense I got is she's very intelligent. You don't get into a position like that without being an intelligent person. I think she thinks she's doing what's best for the world. I don't think she's sitting there like some cartoon supervillain, rubbing her hands together. But they believe equality of outcome should come before all else, at the expense of individual freedom, even reality itself.

> I've noticed that's just kind of the prevailing attitude across all of these tech people I've met in these sorts of positions over the course of my career. They genuinely believe the things they're saying.[25]

Cassandra was clear on what she believed had gone wrong in the tech world and the plans they had for the common man or woman. "I think it's a major group-think kind of mentality. The idea that somehow as a corporate person, you know better than the elected representatives of the American people. There's an attitude that we're better than these other people. Therefore, we should be the ones making the decisions, not the peasants."[26]

* * *

The Project Veritas piece with Gaurav Gite, Jen Gennai, and me, the "anonymous Google whistle-blower," aired, and I knew I'd probably quickly find myself under investigation.

I'd tried some stratagems to avoid suspicion, such as my friendly, cheerful resignation letter, but I expected those two run-ins with Human Resources would quickly flag me as the likely whistleblower. I was certain the information existed in log files **somewhere** inside of Google. Probably hundreds of entries in random access logs deep in the servers that a forensic auditor could find for the past two and a half years as I'd woken up. Ironically, about a third of the content creators that I subscribed to would eventually be purged from the platform in October 2020, right before the elections.

I tried to prepare myself. I would deny it as long as possible, but I'd also start planning for the inevitable confrontation. Some of my friends in whom I'd confided told me Google had a nasty law firm (if you have billions of dollars you can definitely buy some junkyard dog lawyers with Ivy League credentials) that was notoriously mean-spirited in going after individuals whom the company had identified as a target.

I received a letter on July 19, 2019 from Google's attorneys, the law firm of Wilson, Sosini, Goodrich, and Rosati. It arrived both by email and overnight delivery, which indicated to me a fair degree of panic in Google. I include part of it here:

> Dear Mr. Vorhies:
> We represent Google LLC ("Google" or the "Company"), your former employer. In connection with recent discussions with the Company, this letter demands you protect Google's confidential information and also return all

> Google property in your possession within seven days, including the following items in your possession ("Google property"):
>
> 1. One Google badge (bearing the name Zachary Vorhies);
> 2. One GCard; and
> 3. One ITAm Laptop LENOVO X1 YOGA V3 14 2018 US B: 784670: Serial Number: IS20LES30000R90P2J3R.
>
> The letter will also summarize certain of your obligations to Google, and will provide instructions about how to comply with those obligations—including the return of the Google property. In addition, you must cease and desist from any use or disclosure of any internal Google files you possess and provide honest answers, in writing, to each of the questions and demands posed in Section C below no later than Friday, July 26, 2019.[27]

I contacted a lawyer in the San Francisco Bay Area. For purposes of privacy, I'll simply refer to her as "Lady Justice." Lady Justice recommended me to another lawyer who was handling a similar whistleblower case. He told me the case would likely cost several hundreds of thousands of dollars and wanted a large retainer up front. It seemed like the best-case scenario was spending a hundred thousand dollars in attorney fees, which would have bankrupted me at the time. And then there could be a criminal case that would follow after all my money had been spent. It seemed like a bad option, like putting myself into check-mate.

Lady Justice told me I should expect things to get rough, and that Google would try to destroy me. In truth, Lady Justice was tough as nails, a little mean and abrupt, but for some reason she seemed to like me. She gave me some of her time for free, acting as an informal advisor, and told me I could always reach out to her.

But I had one big problem.

I still had Google's laptop and it contained all the evidence of how I accessed the Google network and downloaded the documents. My personal laptop also had copies of the files. Google wanted it, but giving it back to them would inevitably be used by Google to convict me on some trumped up charge, like they had done to the previous whistleblower, Kevin Cernekee.

I had to lose the laptop.

Lady Justice had an idea. She said, "Isn't that criminal evidence, and as such, aren't you obliged to hand that evidence off to law enforcement?"

She was right. The laptop didn't need to be in my hands. It needed to be in the hands of law enforcement. If I didn't hand it over to law enforcement, a case could be made that I was an accomplice to such criminal activity.

I thanked Lady Justice and left.

I went home, put the laptop in a box, along with nine hundred and fifty pages of Google's plans for global-scale censorship. I drafted a letter to law enforcement explaining that the box contained evidence of Google's plans to rig the 2020 election by controlling access to information.

Who did I address the box to? None other than the Department of Justice: Anti-Trust Division.

Google, however, was expecting me to send back the laptop. Instead, I sent them a letter explaining the laptop contained evidence of criminal activity and had been submitted to the Department of Justice. However, I sent the package to a different department, on purpose, because I didn't want Google to intercept it on its five day journey across the country. I figured by the time Google figured it out, the laptop would already be in possession of the Department of Justice.

I still had a number of problems, but a Google laptop wasn't one of them.

A few days later I started to receive a series of comments, suggesting Google was trying to intimidate me.

The first inkling I had that Google was onto me involved a series of comments from this anonymous troll on Twitter, called "1snowflake." (God, I wish I was making that name up! Could there be anything *more* on the nose?) 1snowflake was taunting me on Twitter, referring to me as "Mr. Leaker," and said I needed to change my short bio. It had said I was at Google, but now read I was a "Gainfully employed tech geek in Silicon Valley." Even that wasn't enough for 1snowflake. I was saying to myself, "Holy crap! Who is this guy?"

I called up Patrick, explained what had happened, and asked, "Dude, what's going on?"

Patrick was calming. "Don't respond back," he said. "Let me handle it."

Patrick started interacting with 1snowflake on Twitter, engaging him in an ideological debate, and making fun of some of his arguments. 1snowflake got agitated, arguing with Patrick, and in the course of their discussion, Patrick told him he should check out a certain website to understand why he was wrong.

1snowflake clicked on the link.

Big mistake.

Patrick owned the website. The link immediately logged all the information about the web-browser, giving us the IP address, which we were able to geo-code, and find that it came from a Google data center in Indiana.

I concluded that Google was actively stalking me.

I realized Google was launching a coordinated attack and I had to be prepared to make some moves. Project Veritas had not released all of my documents because that wasn't their style. I realized this was a vulnerability, because although the Project Veritas piece on Google had been devastating and received enormous attention, much was still hidden. I needed to get the target off my back.

I called up James O'Keefe and said, "The Dead Man's switch has been set. I give you permission to release all the Google docs in the event of my untimely death. Can I get you to do that?"

James agreed.

I then went onto Twitter and made a post, knowing they were monitoring me, and wrote, "In the event of my untimely death, all files will become open to the public." I also posted a brutal picture with the tweet, a 1914 black and white image by Franz Stassen, depicting the hero, Siegfried, stabbing the dragon Fafner with a magic sword, from Richard Wagner's opera *Siegfried* (the third of Wagner's four-part Ring Cycle), to drive home my point. In the image, the giant dragon, almost more like a serpent, rears above the hero, as if to terrify him, but in doing so exposes his vulnerable underbelly. The hero, sensing a moment of opportunity, plunges his magic sword, Nothung, into the heart of the great beast. The picture is haunting to me, the dragon with its mouth wide open, claws extended. But the hero's face is calm and resolute, calculating just the right spot for the sword's plunge (this is the picture in the prologue).

How did Google respond to my threat?

Let's review a few events.

On June 19, 2019, I submitted my resignation letter to Google.

One June 24, 2019, Project Veritas released their exposé on Google, featuring me as the "Anonymous Google whistleblower," with my voice changed and my features completely darkened out.

On August 5, 2019, despite the fact I'd resigned nearly two months earlier, Google called the police to perform a "Wellness Check" on me, because they claimed to be worried about my emotional well-being, not because I was a whistleblower.

Who says Big Tech doesn't have a big heart?

* * *

"Dude, these cops. They really want to get their hands on you. And they didn't want to believe you weren't here." It was my best friend, William, who lived in El Cervito, a small suburban community located on the other side of San Francisco Bay.

At my exit interview with Google I'd given them William's address as the place I'd be living. Yes, I'd given Google the wrong address. As Sun Tzu said in *The Art of War*, "All warfare is based on deception." I knew my plan wasn't foolproof, as Google had my current address in their files, as well as the place I'd lived with my old girlfriend when I first left Google to pursue life as an entrepreneur. It wouldn't take much to track me down.

But I knew the police were after me, no doubt at the behest of Google.

I thought that sounded bad. But it wasn't like they had an arrest warrant, right? I mean, a "wellness check?" It sounded voluntary. If I didn't want to talk to them, did I have to?

Fifteen minutes later I heard a BANG, BANG, BANG on my front gate. The building I lived in had a front gate and then a small landing, which led to a few other apartments. It's a common configuration in a typical city.

They started ringing doorbells, and somebody buzzed them in.

The police started banging on my door. I stayed completely quiet. I wracked my brain, trying to think about my legal obligation in the situation. I'd done nothing wrong. There was no "complaint" against me. Google was simply manipulating the police to harass me with a phony "wellness check" and hoping I did something stupid.

The latch on my door was faulty, so after pounding on it a few times, it popped open.

They entered my apartment, calling out that they wanted to talk to me.

But they were scared.

The entrance to my apartment was on the first floor, but my place was on the second floor. You only reached the second floor by ascending a narrow stairway of about thirty-five to forty feet, culminating in a little half turn before you reached the landing. Any law enforcement officer with the slightest bit of training would know it was the perfect set-up for a kill zone. If I'd appeared at the top of the stairs with a gun I'd be shooting right down into them. I could do some serious damage.

I didn't have a firearm, but they didn't know that. After a few minutes, they retreated.

I figured I'd wait them out and then they'd leave. They didn't know I was there. Several minutes passed, and I couldn't hear them talking amongst themselves or to my neighbors. I figured they had left.

I crawled through my place on all fours and went to one of the front windows, which looks down onto Valencia Street. I thought if I put my face in the window, they'd see me. Instead, I turned my phone on camera mode and put it up just over the window sill so I could use it like a submarine periscope. There was a cop car parked across the street. I wondered why it was parked away from my front door, rather than just in front of it.

I crawled away from the windows to my room in the back. As I approached a side window, I again took the precaution to look through the window with my cell phone camera. I saw a police officer on guard outside, on 21st Street, which was a cross street to where I'd seen the police moments before. The police wanted to talk to me so badly that they now had me surrounded.

I though to myself, *Well, they don't have a reason to come in and how do they know I'm here anyway? Maybe I'll just pretend not to be here and they'll go away.*

I crawled on my belly like some James Bond character trying not to trip the laser beam alarms that will lead to his doom. When I'd crawled back to my bedroom, I realized the windows were open. Then I heard a helicopter in the distance.

I realized that the darkest room in my house was my roommate's room, which had no windows facing to the outside world. I crawled back to my roommate's room, and got onto the bed. There was nothing I could do but play dead.

To get my mind off the dread of the situation, I grabbed my roommate's Nintendo Switch console and turned it on. With hand shaking, I selected the game Katana Zero, which was about a rogue swordsman fighting his way through police and gangsters. How appropriate!

Within minutes though, I heard a police helicopter overhead. It circled above, came closer, then further, then closer. Then it was joined by a second helicopter.

There's no way those helicopters are for me, I thought.

But they were.

The police started texting me, saying they wanted to talk to me. I tried ignoring the texts, but they continued. They said they had questions for me, and I replied that they could text me the questions. They answered that they needed to talk to me face to face.

At that point, William drove up, noticing that the streets were blocked off. He fought a sense of rising panic, realizing there was some "event" happening, and I was probably at the center of it.

William found one of the police officers maintaining order and assisting with the evacuations and asked, "Are you trying to find Zach Vorhies?"

"Yeah, who are you?" the policeman replied.

"I'm his best friend, and I'm your best chance of getting him to come out."

The policeman gave William the okay to call me. I saw his number and quickly answered. "Hey, Zach, this is William. I'm outside your place with the police. They're going to stay here. They're not going to leave. They're doing shifts. They just want to ask you some questions for this 'Wellness Test,' to find out if you want to harm yourself."

"Are you kidding me?" I asked.

"Yeah, I know. Totally bogus. But we can solve this quickly if you just come down and answer their questions, face to face."

"Okay, I'll come down and meet them at the gate to answer their questions," I said.

I took a deep breath. Maybe we'd get it worked out. I couldn't believe what Google was doing. They were really concerned about my mental health? Give me a break.

I got up from my bed, walked down my stairs, out the door, and stepped into the small entryway before the gate. When I stepped onto the landing, I saw my way blocked by an enormous, squat bomb robot, almost like a mini-tank, with a long arm attempting to grab the container of denatured alcohol my roommate had left behind. There was a second, smaller bomb robot with moving binocular eyes staring at me.

I called up the police guy in charge. "I can't get out to my gate because of this robot."

"Yeah, we're going to have to wait for it to clear."

"Okay, let me know when it's done, and I'll come back down."

I went back up to my room and waited for them to call. I heard the two helicopters continuing to circle around. Yeah, they were for me. Probably from the local news stations.

Maybe ten minutes passed, and they let me know the robot was clear.

"Okay, I'll meet you at the gate," I said.

"We need to talk to you outside the gate," the officer replied.

I protested. "That wasn't our deal. I said I'd come out to the gate and you'd ask me questions. You're breaking the rules. I'm not meeting." And with that I hung up on him.

I sat in my apartment thinking about the situation. If they could make up an excuse to call the bomb squad, how difficult would it be to make an excuse to barge into my apartment? Their guns would be drawn, maybe I'd get shot, and of course if they were wearing body cameras they'd likely mysteriously malfunction.

I decided I was out of options. I had to do exactly what they wanted, carefully and cautiously.

If I was going to get shot, I wanted it to be in front of a bunch of people, with, I was sure, a lot of them recording the whole event on their cell phones.

I called and let them know I was coming out peacefully.

When I walked out the front gate I had my hands up, but my cellphone was in my hand, recording everything. I saw sharpshooters on the roof across the street, several officers on the sidewalks, all pointing their guns at me, and one officer at the end of the street with a massive looking rifle pointed directly at my chest.

"Lift your hands up higher!" one of the policemen shouted at me.

I'd made the mistake of wearing my jacket and immediately realized they thought I might be wearing a suicide belt of explosives. I'm a skinny guy, so when I reached my arms as high as they could go, it lifted up my jacket and shirt so they could all see my pale white belly.

Maybe it was just my imagination, but I felt the tension seem to ease in the officers. *Be compliant, Zach,* I told myself. *Be compliant.*

"Okay, turn around now, keeping your arms up," said the officer with the big rifle.

I did as I was asked.

"Now, keeping your arms up, start walking backward toward us."

That's pretty smart, I thought. *They know I don't have a suicide vest, I'm turned around, and I'm walking toward them. They should be feeling pretty safe. I might not get shot.*

They had me walk about a hundred feet down the sidewalk, passing a small convenience store at the corner of the street with a concealed alcove. I couldn't see them, but three policemen were hiding there, armed to the teeth.

When I passed the store, they burst out of hiding, grabbing my arms and putting them behind my back. They searched me, not finding weapons, and then walked me over to the main officer, the one with the rifle. Thankfully, it was no longer pointed at me.

He introduced himself, and then said he wanted to ask me the wellness questions. I nodded.

"Have you been having any thoughts of harming yourself?"

"No."

"Have you been having any thoughts of harming somebody else?"

"No."

He started to look a little confused. "Have you been having any thoughts of harming yourself or others?"

"No."

"Have you missed any medications?"

"No."

Now he was really confused. "Well, do you know why Google would call in a 'wellness check' on you?"

"Probably because I'm blowing the whistle on all their illegal activity."

The officer fixed me with a curious look. "What?"

"I can prove it."

I got out my phone. I'd taken a picture of the letter I'd written to the Department of Justice when I'd turned over the 950 pages of documents, and actually read some of the letter out loud.

The officer looked both relieved and amused, saying, "Okay, I guess his story checks out," and the remaining police started treating me like a good guy. I heard one of them mutter under his breath, "Fucking Google," as the tension completely drained out of the situation.

The boys in blue didn't have much love for the tech lords of Silicon Valley.

The officer spoke into his walkie-talkie, letting everybody know the situation was over, and in a few minutes the two news helicopters circling above got the message and flew away. I was feeling good because these were the kind of all-American guys for whom I'd become a whistle-blower. We were on the same side. They didn't have an insider in the tech industry to let them know what was really happening. They expected to simply be able to turn on their computers and learn what was going on in the world, never imagining they were being lied to in a seamless fashion. I liked these guys who put their lives on the line every single day and started to joke around with them.

"Hey, I really love cop shows, but I never know which one is the most realistic. *Law and Order*, *CSI*, some other one?"

There was a pause from the circle of police and then one of them said, "*Reno 911*," and the others started laughing. For those who haven't watched the series, it's a mockumentary-style parody of a typical cop show, trying to look like the long-running gritty *Cops* reality TV show, but the police are all a bunch of incompetent, self-absorbed, clueless boobs.

I said, "No, I'm being serious," but that only made them laugh harder. There was universal agreement that *Reno 911* was the most realistic cop show around.

I guess they were telling me it was a *Reno 911* world out there, and I needed to take care of myself. The police packed up, the residents, customers, and owners got back to their places, and I went back to my apartment.

I was done with all of the sneaking around, trying to stay anonymous crap.

James O'Keefe had told me I'd be safer if I went public with my true identity, but I'd questioned his motives. But he was completely right. If I was holding documents back, Google still had something to gain. Trying to intimidate me into silence was a reasonable play for them.

I needed to change the game. By activating the Deadman's switch and letting the world know my name, I was giving myself the best chance of survival.

I needed to get out of San Francisco, record another interview with Project Veritas, and reveal myself to the world.

CHAPTER TWELVE

Reclaiming My Soul

The next day I booked a ticket to Washington, DC where I met up with Patrick and a friend of his, Connie Elliot, a brilliant and vivacious woman with a number of media contacts. Together we planned what came next.

They picked me up at the airport and as soon as I got in the car, Connie started talking. She was delightful. Connie is half Vietnamese and half-European, and a hundred percent anti-communist. Her mother had a harrowing escape from communist Vietnam and her father writes books on how civil wars happen in countries.

Connie provided an endless stream of commentary, and also paid for a hotel room for me near where she was staying, so I couldn't be tracked with my credit card. We realized we were in a very precarious danger zone, with Google knowing they'd blown their chance to take me off the chess board, and wondering what I'd do next.

After I was safely in my hotel room, Connie started going through her media contacts.

The journalist, Sara Carter, well-known because of her many appearances on *Fox News,* was interested in the story, and Mike Moore from *True Pundit,* also showed interest in my story. It didn't seem like there would be any trouble getting media attention. We told them we'd give them the story, but only after I'd come out with Project Veritas.

One of the important things I needed to do in Washington, DC was to make sure the 950 pages of documents had made it to the right person at the Department of Justice, Anti-Trust Division. We made some phone calls to make sure the evidence was being processed correctly. However, I quickly realized I'd made a fundamental mistake. I didn't address my package to anyone specific. The evidence hadn't quite vanished into a black hole, but let's just say it was circling one and it would probably be several months before anybody looked at it.

We got in contact with a judicial watch group called Groundswell, and they put us in touch with a specific person at the Department of Justice, to whom we should address the materials. For the second time, I made a submission to the Department of Justice and hoped that the best eyes would be looking at the Google documents.

* * *

The next day Patrick and I went to New York and the Project Veritas headquarters. James and the rest of his team greeted us warmly, and within a short period of time I was sitting in the interview chair, with James across from me, and the crew ready to film. It was an emotional interview and at one point I started to tear up. Here's the first minute of the nineteen-minute video that Project Veritas put together and released on August 14, 2019. I was highly emotional and hoped I'd make sense. Sometimes my thoughts could be jumbled, but I trusted that people would understand as they watched the entire nineteen-minute video.

> **Zach:** For three years, since 2016, when they started changing everything, and to have that burden lifted off my soul, is, I've never felt happier, or more at peace with myself than I have right now.[1]

With that opening, they showed footage of me on Valencia Street in San Francisco, surrendering to the police, hands over my head, cell-phone in my right hand, police officers crowding the frame, telling me to turn around and walk toward them. The video then switched back to James and me in the studio.

> **Zach:** The police didn't announce themselves. I decided, I'm just not going to talk to them.
>
> **James:** So, why are the police outside your home in San Francisco?
>
> **Zach:** What I did is I put out a Dead Man's switch. People say I'm brave when I'm explaining what Google is doing. It's an act of atonement. To make my conscience clear.[2]

Project Veritas tried to portray me as the "Google 'Machine Learning Fairness' Whistleblower," and I thought that was good framing. I considered the "Machine Learning Fairness" program to be the greatest threat to our system of democracy and understanding of reality. However, I know it's not that catchy.

In the interview, I talked about what had happened with Google, the way certain news was being suppressed, and that other globalist friendly news was being amplified. I felt I was essentially telling the American public something many already suspected. But I was giving them the details of how things were being changed.

It's a surreal, out of body experience to be interviewed, and you're not certain if you're making sense, or sounding like a complete idiot. When I finished the interview, I looked at the crew for their reaction. They told me it was amazing, and I accepted I did an okay job. I felt exhausted, but also invigorated.

The story went into production and I was just hanging around the Project Veritas office. At one point James said they needed some pickup shots of me outside in nature, walking along with him and talking. While I was filming with James, my cellphone rang, and I answered it. It was the lawyers for Google, and they wanted to talk to me.

I told them I was busy and said I'd call them back.

Of course, I had no intention of calling them. I turned my phone to vibrate mode and put it back in my pocket. I wasn't answering my phone until after the video dropped. I genuinely believed that if the disclosure didn't come out, I would likely be targeted for assassination, dying in some mysterious "accident." The lawyers kept calling, and I refused to answer them. There was no sense in it. I wasn't backing down.

The afternoon before the segment was to air, James O'Keefe came up to us and said, "I want to show you something. It's gonna take a few hours. You up for a trip?"

The three of us piled into James' car and drove to Long Island Sound, where James kept his sailboat. At the dock we took a small, motorized craft to his sailboat which was moored out in the water. James seemed at home on the boat, quickly checking things, and the pressures of running Project Veritas fell away from him as we took in the clean Atlantic air and felt the calm the ocean can bring to a person. We'd picked up some sandwiches and beer along the way and ate in the fading summer light. After we'd eaten, James brought out some cigars, and we lit them up.

James turned philosophical at that point, and I felt honored by his attention. He wanted to talk about what was going to happen the following day, and what it meant for society. We were the ones who knew the full extent of Google's totalitarian plot and we were the ones making progress to expose and reverse it.

He also wanted to reassure me that coming out was the right thing to do. He said it was his experience with whistleblowers that once they came out publicly the companies lost interest and didn't prosecute their claims. Why would they? It only gave more coverage to the claims of the whistleblower. I hadn't believed him before when he'd made that argument, and it had nearly gotten me killed.

James said you needed to shine a light on these criminal acts and when you did those committing them would try to run away like cockroaches. He talked about using the power of the First Amendment to expose these people and set back their plans. James was aware we had a big task in front of us, but said we also had a great responsibility to inform the public. He was passionate that the framers of the Constitution were wise in making freedom of speech the First Amendment, because all our other freedoms flowed from that principle.

It was a great night, a special night, and I felt an overwhelming sense of calm settle over me. For the last two and a half years I felt as if I'd been going through hell, coming to understand what Google was attempting, and being complicit in that system by accepting a paycheck from them. I was no longer going to suffer in silence. Instead, I'd stand up and say what I knew was true and deal with the ramifications of my decision. I felt I had a duty to be a public figure and show people it was okay for somebody on the inside to step forward and reveal what was happening inside Big Tech.

My hope was that other whistleblowers would follow in my footsteps. I thought of the motto of Project Veritas, "Be brave. Do something." I hoped people would think of that motto one day as they thought about me, that I had been brave. I also realized I had an enormous obligation to be a good role model with my every move being watched and my every utterance closely scrutinized.

And yet I was ready for it all. Since the ordeal had started, I'd often wondered whether I was up to the challenge. Would I show myself to be a person of integrity, or would I take the coward's way out and maintain my silence?

I slept more deeply that night than I had for several months and woke without fear of what the future might bring. Whatever happened, I would accept it without complaint. I was living my life as a free man, on my own terms, and telling my truth.

* * *

BOOM!

On August 14, 2019, Project Veritas published their nineteen minute and thirty-four second video report on my story.[3]

It was a bombshell, confirming what some people had long suspected about Google and their attempt to manipulate the news. And for the first time in their history, Project Veritas did a document dump of all 952 pages I'd downloaded from Google. Project Veritas had a link to the story, so they'd get an email notification every time somebody downloaded the documents. I was watching their email feed and they were getting multiple emails a second and thousands of times an hour.

I'd dumped a treasure trove for data geeks around the world and they were devouring it with glee, even picking up things I'd overlooked. In looking over the YouTube blacklist, many noted that nearly all of their terms resulted in suppressing searches for "false flag" events. Others were asking why YouTube was suppressing searches for CIA tradecraft and assassination techniques, the Eighth Amendment to the Constitution of Ireland (recognizing the equal right of the unborn child and its mother), as well as supposed cancer cures using natural products. All the claims the conspiracy theorists had been making about Google suppression of information had suddenly become conspiracy fact.

I imagined the Google attorneys shitting their pants and my former co-workers being horrified to discover that the person who sat a few cubicles away from them had given information to the hated Project Veritas. And yet I feel it's important to give a message to my former colleagues still behind the lines in Silicon Valley. I understand the buccaneer, break the rules mentality of the tech industry, and realize those qualities are more likely to create a person with liberal leanings. I dislike when I hear conservatives write off tech people as irredeemably liberal, and thus certain to engage in collectivist ideologies as pursued in the former Soviet Union and China. I do not think such an outcome is inevitable. My liberalism, and that of many others, has more of a 1970s flavor, skeptical of big business and the military/industrial complex, as well as a government that wants to tell me how to live my life. However, I do clearly see the danger that unchallenged liberalism can lead to authoritarianism.

And yet I believe the psychological operating system of those in the tech world is a rugged individualism, which is at odds with such a system of control. The video games so many of us played as children featured a lone hero up against tremendous odds. I believe these games gave us a mental map for fighting an all-powerful system of control, and the fact that if we died in the game we simply started over again, gave us the confidence that even if many attempts failed, there was a path to victory. I believe without reservation that many of my former colleagues in Silicon Valley watched the Project Veritas release and started to question what their bosses were really doing.

Because we had contacted many various media outlets and let them know what was coming, I did six interviews on the day of the Project Veritas release. I was about to become a significant media personality.

For the next several months I was on the interview circuit, averaging about two a day. Thank God for the four years in high school I spent in speech and debate. I kept telling the same story, over and over, eventually developing my fifteen-minute version, half-hour account, and my hour long one, depending on the format of the interview.

While the interviews generally went well, I did make one mistake. I was uncertain whether it was a good idea to talk about the YouTube shooting I'd observed, thinking that the issue of whether the shooter was a man or a woman would distract from my story. In retrospect, I think it was a mistake for me not to talk about that incident.

The tech geeks were quickly dissecting my history, noting my appearance on *America's Greatest Makers*, but also wondering why I hadn't talked about the YouTube shooting. They started questioning whether I was really going after Big Tech, or whether I was some sophisticated Deep State counter-intelligence actor. I'm reminded of the quip from President Reagan that "sometimes the right hand doesn't know what the far-right hand is doing." I am not some Deep State actor, and the paltry amount of money in my bank account surely supports that claim.

In October 2019 I was invited to the American Priorities Conference (AMPFest) held at the Trump Doral Country Club in Florida to be a featured speaker about my experiences. I started meeting a lot of my long-time heroes like Tom Fitton of Judicial Watch, or the investigative journalist, Tracy Beanz.

For the first time, I felt like a celebrity, because most of the people had seen the Project Veritas video, and I had the experience of walking by people and hearing them whisper, "Oh my God, is that Zach?" People asked me to take pictures with them and they'd tell me I was brave and ask questions about what I thought would happen next. I never wanted to be a celebrity; I just wanted to expose the truth.

At my talk, I showed a series of slides, detailing the Google plan for complete information dominance, and when it was over, I got two standing ovations. It's difficult for me to put into words how emotional it was for me to get such a response. I'd been in a very dark place, alone, isolated, and outnumbered by some extremely dangerous people. But it was an illusion, generated to maintain compliance. This was the real world, the average man and woman, just wanting to get accurate information without bias, willing to listen to different points of view, and quite capable of coming to their own conclusions.

* * *

During those hectic months, I'd often get a question near the end of the interview about what I saw for the future. It wasn't necessarily about Google, but where we were heading as a society. Could I provide any insight?

It had been an interesting evolution for me, going from being a foot soldier in the Occupy Wall Street movement to a featured speaker at the American Priorities Conference at the Trump Doral Country Club.

My answer is that we're at a critical point in human history.

For several thousands of years societies were built on the value of human and animal capital, slaves to build the pyramids, or cattle to plow the fields. As we developed machines which could do those tasks, people became more physically free. With the invention of the Guttenberg printing press more than five hundred years ago, we could become more intellectually free. No longer did the people need to depend upon an elite caste of priests who could read the relatively few and precious copies of the Bible, available only in Latin. The Protestant Reformer, Martin Luther, translated and printed the Bible into his native German, so it could be read and interpreted by the common person.

Amazing progress followed in the ensuing centuries.

And yet these new technologies also brought with them enormous dangers.

With the advent of the twentieth century and the marvels of radio, photography, and moving pictures, the combustion engine, airplanes, and cars, the world was poised for an unparalleled golden age of progress. But the tyrants among us, people like Lenin, Hitler, and Stalin, realized that if they could control these items, starting with the media, then they could take us in a much darker direction. What might be a force for progress and efficiency, could just as easily be turned into an instrument of evil.

I believe our technology, particularly artificial intelligence, poses just such a dilemma for us in the modern world. When I'm driving to meet a friend, using Google maps from my cellphone to find his new place, and the computer voice comes over my speakers (interrupting the audio book which I'm also listening to on my phone) and tells me there's an accident ahead which is going to make me ten minutes late, I can quickly call my friend and let him know. In that simple interaction, consider how many ways my cellphone is a benefit to me.

First, I'm not getting lost.

Second, I'm listening to a book as I drive.

Third, I'm learning about an accident before I encounter it, as well as getting an estimate for how late it's going to make me.

The information architecture of social media was designed to be a simple feedback loop, allowing consumers to participate in the experience, providing them access to desired content so they might more fully develop themselves in the way they want. When one considers that the old model was one of top down control, where we were told what to believe and what was true, and then reinforced by our religious institutions, our print, radio, television news, and social media offered an unparalleled expansion of direct democracy. The old, legacy media was somewhat in the dark as to how they were being received by the public, so for many issues they tried to give us a relatively accurate view of the world, as best they understood it.

At the dawn of the social media age, we saw a remarkable expansion of true democracy. If you wanted to educate yourself about a subject, you could easily find just about everything ever written about it.

But like the tyrants of old, the avatars of this new technological era saw how these new technologies might be turned to serve their ends. If you knew what people were looking for, you could appear to give it to them, but it would be what you wanted them to know and believe.

The question would be if they could get away with it.

The wild card in this equation is the human being.

Consider this scenario. Your daughter says she wants a dog for her birthday. When her birthday arrives, you give her a hamster. Her first response is to break down in tears.

You tell her a hamster is just as good as a dog.

It doesn't work.

Next, you resort to outright lying. Maybe you tell her scientists have discovered that dogs can carry a deadly disease, and if she gets a dog, it might kill her favorite grandmother.

She does a little research on her own and finds out that you're lying.

The lies and arguing continue for years. Maybe you tell her even bigger lies. Perhaps there are even some good reasons you don't want a dog, but you've never shared them with her. Maybe when you were a kid you were attacked by a German Shepherd or pit bull and you have this deep, irrational fear of all dogs in the pit of your stomach, but you can't even verbalize it to her.

The simple fact is you have lost your daughter's trust.

Once broken, trust is extremely hard to re-establish. Normally, it requires nothing less than complete honesty.

If the father wants to restore his daughter's trust, he must tell her the truth. In the same way, we must know the truth. I foresee two possible futures.

In the first possible future, we accept what the artificial intelligence tells us. We wanted a dog, but calmly accept that a hamster is just as good of a pet because that's what Google or YouTube, or whatever system of control then in place, tells us. And besides, we wouldn't want our dog to kill grandma. In this scenario, we become slaves to the system.

In the second possible future, we rebel. Considering the long history of humanity's aggression, violence, and warfare, I believe this to be the more likely outcome. Put simply, we're not really a species who likes to behave. If that was the case, we wouldn't have so many alcoholics, drug abusers, philanderers, or criminals. I agree that most people in the world like to go along with the prevailing social customs. And it may appear for a time that the powers of conformity have won. But there is an eternal schism in man's nature, between compliance and rebellion. And neither side can ever be completely extinguished.

The challenge is how to live in harmony with both sides.

Whatever you try to repress will come back so much stronger.

Tell somebody they can't think a certain way, and they will choose that as the hill upon which they will die. The specific issue doesn't even really matter. We need freedom the same way we need air, food, and water.

By offering up our lives in defense of freedom, we preserve it for those who come after us. The artificial intelligence will be our servant, not our master.

I believe we're heading for what he calls the "Great Technological Schism," akin to the great religious schism in Europe of the Middle Ages between the Catholic Church and Protestantism. The system of software a country uses will reflect its internal cultural imperatives and its unique cultural identity.

The elites wanted to have a unified social media system, whether it was Google, Facebook, YouTube, or Twitter, and wanted to make these programs ubiquitous around the world.

But it didn't work out as they hoped it would. For example, the revolution in Egypt which brought the Muslim Brotherhood to power was started by a former Google employee, Wael Ghonim, who later wrote a book called *Revolution 2.0: The Power of the People is Greater than the People in Power.* This is from a 2013 article on the book and his campaign in Egypt:

> Written by the now-famous Google executive and accidental revolutionary Wael Ghonim, the book provides you with an amazing bird's eye view of the build-up to the uprising and its successful conversion into a force for democracy, social justice and respect for human rights—or, to use the inspiring revolutionary chant, "bread, freedom, and human dignity.[4]

Sounds like a wonderful revolution brought to you by Big Tech, right? Let's review how it was later described in a long article by the Carnegie Endowment for International Peace. At first, the Muslim Brotherhood, aided by former Google employee Wael Ghonim, and a Facebook page which attracted great attention, seemed to make all the right moves.

> Throughout the eighteen days of demonstrations in January and February 2011 that toppled Mubarak, the Brotherhood was careful not to be perceived as taking control of the protest movement in terms of its slogans, discourse, or political demands. Brotherhood leaders were aware that the protests were not dominated by Islamist ideas but rather oriented toward the broad goals of freedom and social justice.
>
> They were also aware that other political groups were instrumental in mobilizing demonstrators and writing the narrative of the uprising.[5]

Sounds like the Muslim Brotherhood let all the other groups do the difficult work and then simply hung around to take power when the other groups started fighting with each other. It's one thing to start a revolution (or perhaps more accurately, take one over), but it's another to turn it into a stable government. Trying to govern effectively is where the Muslim Brotherhood completely failed.

> From early 2011 to the middle of 2013, Egypt's Muslim Brotherhood failed to lead an inclusive democratic transition, appreciate the full diversity of

> Egyptian society, and understand the need for a completely reinvented political culture . . .
>
> For the political inclusion of the Brotherhood to lead to the group's democratization, two conditions were necessary. First, post-Mubarak Egypt required a consensus on new rules of the political game. Second, the Brotherhood needed to undergo an ideological and organizational transformation, including by embracing the principles of democracy, pluralism, individual freedoms, citizenship, and equality before the law.[6]

The Carnegie Endowment lays it all out in a clear fashion. The Muslim Brotherhood did not lead the democratic revolution that overthrew Mubarak. But in the aftermath, they did take control of the government. Since they'd cloaked their true extremist views during the revolution, it's not surprising they weren't able to win over a public which was interested in the political freedoms more commonly found in western democracies.

And what was the role of Big Tech in all of this? The answer is murky at best. But since we have a former Google engineer leading the revolution, helped along with his Facebook page, I couldn't help but draw parallels with my Occupy Wall Street activities and the subsequent infiltration by Antifa. It's said that the human mind inevitably finds patterns, even when none exist, such as figures in the clouds. But sometimes there are patterns and we're foolish if we don't see them. If I put an egg in a pot of boiling water, I'll have a hard-boiled egg in a few minutes, every time.

After the disaster of the Muslim Brotherhood in Egypt, other countries started moving away from the Big Tech platforms of Silicon Valley. A software program called Project Jigsaw allowed countries like Syria and Libya to begin building their own search engines similar to Google, but under the control of those governments. Russia did something similar with its Yandex system, which allowed it to create its own social media platforms.

I think people will continue to move away from these Silicon Valley tech platforms and the original promise of the internet can be fulfilled, allowing all the people of the world to meet each other in conversation and discussion. In the past I believe our "meta-consciousness" was driven by the top down broadcast model of newspapers, radio, and television. These institutions, relatively small in comparison to many industries, were open to capture by the reigning oligarchs, giving the writers an option to either write the story the way those in power wanted it, or finding another line of work.

But today's social media has the potential to create a bottom-up meta consciousness, where a question is asked, the people respond, and the best responses float to the top to acquire the greatest amount of support. I genuinely believe we're breaking the old system of top down bureaucratic broadcast communication. We're birthing a symmetrical, bottom to top feedback social media system where the common people can come out and tell these social media personalities they've got things wrong. As long as we keep the "marketplace of ideas" open and healthy, just the way we try to do with anti-trust laws, none of the players will be able to gain a dominant position to squelch the opposing voices.

I think it's going to be a great future.

It just might be a little bumpy before we get there.

CHAPTER THIRTEEN

What Bravery Brings to You

I thought my story was over.

What else was left? I'd publicly revealed myself, been celebrated at the American Priorities Conference at the Trump Doral Country Club, and people were calling me a hero. That should have been the high-water mark. I expected things to slow down in my life.

In truth, I was just getting started.

* * *

Somebody who would become very important to me was deeply moved by my Project Veritas interview. Maryam Henein, a brilliant and beautiful investigative journalist, who'd directed the documentary *The Vanishing of the Bees*, in 2009, narrated by the well-known actor Elliot Page (then still known professionally as Ellen Page). The film detailed how Bayer's pesticides were causing the collapse of honeybee colonies in North America and Europe, with the implicit message that we were also endangering our own health.

Maryam watched the interview, looked into my eyes, and felt as if she already knew me. She'd been doing her own research on Google, because their information crackdown on natural heath websites had caused a 76 percent plunge in web-traffic to her own health site, Honey Colony. She reached out to me on Twitter and asked for an interview. She hoped I could come to Los Angeles for the interview in person. It just so happened that the next week I was going to be in southern California visiting with my mentor, Patrick Ryan.

I said that before I'd do the interview I wanted to spend some time talking to her privately to be certain she was legitimate. Maryam looks like a model, tall and willowy, with Egyptian features and a French sense of style since she grew up in Montreal, Quebec, Canada. The French-Canadian accent makes her seem more mysterious and more than one person who's met her has questioned whether she's a spy for some intelligence service. Half my brain was flipping out over her and the other half was wary. Was she a honey-trap?

But as I talked with her and learned her history of activism, I thought, *Man, if she's a honey-trap, she's got the best fake identity I could ever imagine.* The three of us hung out for a while.

Maryam suggested we go for a walk around Lake Hollywood, near her house. It's a reservoir with a walking trail of about three miles, and if you just focused on the water and surrounding area you might think it was some Alpine lake in the Sierra Nevada Mountains on the California/Nevada border. The number of things we had in common quickly became clear. We both hated Google, were big supporters of natural health, believed the government was lying about the harmful effects of vaccines and other products, and wanted to fight for the future of America.

A month after we met, I asked if I could kidnap her to come live with me in San Francisco. She said yes. Maryam came to stay with me in San Francisco in October and November 2019, and in January 2020 she got rid of her place in the Hollywood hills and moved in with me.

A few weeks after she arrived, we started to get word of COVID-19, the coronavirus which originated in Wuhan, China, and started to see videos about it. We started covering the early days of the outbreak for Alex Jones on *Infowars*, and for those first few weeks we were genuinely terrified. We got the masks, the alcohol wipes, and even got our own Hazmat suits. If it *was* the apocalypse, we were going to be ready.

For several years, Maryam had been drawn to Costa Rica, so as the COVID-19 madness began to build she asked me, "Can I kidnap you to Costa Rica?"

It's probably not surprising, considering how many things Maryam and I seemed to be in synch with, that I'd also been interested in Costa Rica for some time. "Sure," I said.

We planned to stay for two weeks but ended up landing in Costa Rica in March 2020 on the same day that COVID-19 supposedly landed in the country.

Instead of being there for two weeks, we ended up staying for four months. The jungle wasn't a bad place to ride out the viral apocalypse.

* * *

One of the people I'd been introduced to after coming out with the Project Veritas video was the filmmaker Mikki Willis, who'd been the official videographer for the Bernie Sanders presidential campaign of 2016. By 2020, Mikki had turned away from the socialist Sanders and become libertarian, and Mikki was fascinated with my story of big tech censorship. One of the enormous surprises of 2020 was the amazing success of a book called *Plague of Corruption*, by Dr. Judy Mikovits, a longtime government researcher, and Kent Heckenlively, a science teacher and attorney (and also my co-author). Mikki lived near Dr. Mikovits and knew her story and he put together a twenty-two-minute interview with Mikovits called *Plandemic*, which, despite being banned by most of the social media platforms, eventually reached mega-viral status. Judy's coauthor, Kent, had written the opening narration. Based on the boost provided by Mikki's interview, the book eventually reached number one of all books on Amazon for a week, beating out Michelle Obama's book, *Becoming*, and those wimpy teen vampires in the latest installment of the *Twilight* saga. The book would also spend six weeks on the *New York Times* bestseller list.

In the wake of the controversy over *Plandemic*, Kent discovered a phony Twitter account for Dr. Mikovits and when Mikki learned about it, he contacted me. I said the only way to fight it was for Dr. Mikovits to create her own Twitter account. Mikovits agreed, I set up an account for her, and she quickly zoomed to more than seventy thousand followers in the first week. I started managing her account for her.

But then the problems began.

One of the issues which often arise in activist communities is there are a lot of competing voices, and sometimes it can quickly turn into a mess. If you're a counter-thinker, seeing corruption where others see enlightened authority, it can often bleed over into your personal interactions. People with rebellious natures rarely make the best collaborators. Mikovits had some people around her in Southern California who didn't appreciate how you needed to message social media and said she should do it on her own. (This would eventually become a moot point as Twitter would ban her from posting, but I was getting pushed aside, and it rankled me.) Mikki thought Kent, who lived in Northern California, should be brought in to troubleshoot the problem.

It quickly became clear to me Kent was a calm, rational person, trying to balance everybody's interests. The situation wasn't ideal, but he showed himself to be a straight shooter. He also didn't have any trouble telling Judy she should continue to use my services. But she said she wanted to handle her own social media. It's a rare person who can disagree without being disagreeable and not lose friends on either side of the argument. And it was also apparent Kent was the mastermind behind the book's accessible storyline and easy to understand science, and he was aware of the need for help from others, like Mikki, to help the book become a bestseller.

After we'd finally settled the issue of my social media work for Judy, I asked Kent if he wanted to help write my story. There was a pause on the line and then he said, "Zach, since we're on the phone, you can't see me. But right now, I'm down on my knees, begging you to let me help write your story."

As with Maryam, I'd come to find those interested in natural health, as well as being concerned about vaccines, were often the quickest to understand my story of Big Tech censorship and manipulation. They often considered themselves the first victims of Big Tech, since they'd originally created their online communities using Google and Facebook, only to find themselves under increasing attack and outright banning. In 2017, Kent himself had been publicly banned for three years from the entire continent of Australia, because he wanted to talk about corruption in vaccine science. He took it with characteristic good humor, noting the only other people who'd been publicly banned by Australia were rappers who viciously beat up their girlfriends, alleged white supremacists, and people like Chelsea Manning, who'd stolen military secrets. None of these categories applied to Kent, a mild-mannered middle school science teacher, who simply wrote books Big Pharma (and apparently their paid minions in government) didn't like.

From the jungles of Costa Rica, Kent began taping the interviews with me that became the basis of this book. When I returned to San Francisco in June 2020, Kent and I made plans to meet, get lunch, and spend a few hours together, as he wanted me to show him several of the physical locations where the events of my story took place, such as the "wellness check" I'd been forced to endure on Valencia Street. In person, Kent reminded me of the actor Matt Damon, an intelligent, self-effacing, All-American dad next door type with a hint of mischief in his eyes. "I figured you'd want to meet me in person to make sure I'm not some Deep State actor," he said when we sat down for lunch. "Ask me anything you want."

"I figure you're okay," I said. "I checked you out." Like Maryam, he had a long history of activism, and in addition, a daughter with severe autism.

It would be a lot to fake. "If you do have intelligence connections, it seems like you'd be hooked up with the good guys."

He laughed. "I'm pretty much ignored by all the groups," he said.

"That makes it more likely you're an independent actor. If you had intelligence behind you, they'd provide a group for cover."

"Now I know why nobody comes out to play with me!" he replied.

We had a good lunch and visit, and he was an excellent listener. By that time, we'd already recorded several interviews, and I felt like I already knew him. He had that quality of paying attention, as if you were telling him the most important story in the whole world. If not the most important, I certainly thought it ranked in the top ten.

* * *

I was invited back to the American Priorities Conference (Amp-Fest) at the Trump Doral Country Club in October 2020, a few weeks before the presidential election. Not only did they want me to be a main speaker, but they wanted me to put together three panels on censorship.

Each session would be an hour and a half, so I proposed we have one session on "How Big Tech Censorship Works," one on "Health Censorship," and another called "Stop Bit Burning." Bit burning describes when the search engines simply de-list certain websites or information, a process which many have likened to the Nazi practice of burning controversial books. For several months I'd been part of a group which had been trying to bring a class action to get the practice stopped by Google and all tech companies.

If you've never been part of planning a large conference, let me give you a little advice. Expect things to change radically, from day to day. Originally, I figured I'd be responsible for four and a half hours of talks at Amp-Fest. A little daunting, right? The "Health Censorship" panel was the first to fall, as I wanted to talk mainly about misinformation about vaccines, only to find out there was somebody else doing the very same topic. And I was having trouble finding people for that session, as Robert Kennedy, Jr. was having some elective surgery done, and the filmmaker, Mikki Willis, was also too busy to attend.

The "Stop Bit Burning" panel quickly ran into trouble when it turned out one of my main speakers was greatly disliked by the festival promoters as he'd had some sharp criticisms of the conference over the years. The "Stop Bit Burning" panel also got yanked.

But that left me with an hour and a half to fill with the "How Big Tech Censorship Works" panel. However, as I started working on that I got a call from the festival organizers telling me my panel was cut from an hour and a half to thirty minutes. I went from being worried about not having enough people and content to wondering how I was going to fit all the information and people into the time allotted. The two people I'd picked for the panel were Ryan Hartwig and Zach McElroy (yes, another Zach!), two Facebook whistle-blowers featured by Project Veritas. Both Ryan and Zach worked for Cognizant, which provided content moderation for Facebook. Do I give away the surprise if I tell you Facebook was biased against conservatives in general and Donald Trump in particular?

I called up Ryan and Zach and said, "Hey guys, we've got thirty minutes, so we really have to make it quick. I've got an opening planned and a question period, so get your talk down to seven minutes."

Zach quickly got his talk down to seven minutes, but Ryan had more trouble. He had a presentation of more than a hundred slides and there was no way he was going to get through them in seven minutes. I sympathized with him greatly. When you come out as a whistleblower, you want people to know you have lots of information, and you want to give it all to them. But if you try to give it all to them, you will drown them in detail. The trick is to pick out the most shocking claims, present them first, make it relevant to them, and then backfill with detail.

When I got together with Ryan and Zach at Amp-Fest to work on the presentation, I said to Ryan, "What's the most shocking thing Facebook is doing? You need to lead with that."

Ryan thought for a moment and then showed me some pictures Facebook allowed. Despite their prohibition on allowing images of violence, Facebook allowed pictures of Donald Trump having his neck sliced with a knife or having his brains blown out with a gun.

"That's great," I said. "The important thing about this story is not you, Ryan Hartwig, but the people in the audience getting censored. If you make it about the audience, instead of being about you, then you have them."

Next, I had to work on my presentation. I'd planned to talk about Google using Wikipedia to page rank the entire internet. I wanted people to understand how Wikipedia was at the center of so much of the evil crap that Google was doing. But there wasn't enough time. And in addition, I knew I had to capture the audience's attention right at the start.

I put together a short video to open our talk, using drone footage of San Francisco taken during the crazy wildfires of 2020, when the city was choked with so much smoke it looked almost orange with the pollution like some future dystopia. Then I cut to Facebook CEO Mark Zuckerberg saying Facebook is not going to let anybody declare there's a winner until there's a "consensus result." I had some hip science fiction music playing like one might find from the classic 1980s film *Blade Runner.* As the video finished, I stepped out onto the stage wearing the mask of light that I'd built for the reality TV show, *America's Greatest Makers.*

I'd reprogrammed the LED lights on the mask to play an American flag waving in the wind as I walked slowly to the podium. I pulled off my mask and said, "My name is Zach, and I'm from the future." The audience exploded with applause, and I waited for them to calm down before I continued. "In the future the communist cabal has been defeated. Prosperity has returned to the American people. And Donald Trump is the president of the United States."

The audience went crazy and I knew we were off to a good start. After they settled, I continued. "But today, to save the future, we still have lots to do. Facebook is censoring millions of Americans. And to tell you more about that I present Ryan Hartwig and Zach McElroy, the Facebook whistleblowers. Put your hands together now for Ryan Hartwig."

Ryan took the stage and the next image the audience saw was a picture of Donald Trump having his head cut off with a knife. Ryan gave an excellent speech, as did the other Zach. The message was clear. Violence against conservatives was allowed by Facebook, meaning that Big Tech had declared open season on a good chunk of the American population.

We had a brief question and answer session with the audience, and after it was over, I was approached by another big tech whistleblower. I am intentionally not naming him at this time, but I can vouch for his credibility. He told me the reason lawsuits against the tech giants were failing was because of a secret agreement with the National Security Agency (NSA). In exchange for handing over all their data to the NSA, Google, Facebook, Twitter, and others were given complete immunity from lawsuits by citizens alleging censorship.

Section 230 of the Federal Communications Act gave big tech freedom from lawsuits because they are not technically "publishers." It was being challenged in several court cases because so many of the platforms are alleged to be acting as "publishers." However, this was section 230 on steroids, giving the government unprecedented information on its citizens, while big tech escaped scot free. I know my friends in the health freedom movement would immediately recognize the move, as it was the same thing which happened in 1986 when vaccine makers were given complete immunity for harm to children caused by their vaccines. How could any sane person believe corruption would not take place if you let powerful interests operate without any responsibility for their actions?

I thought it was important information to be shared with the audience, so I stepped back on the stage, formulating my thoughts as I asked again for their attention. I explained what I'd been told and then suggested interested groups might want to start filing Freedom of Information (FOIA) requests to see if we could pry that information loose from the intelligence services.

If the information I'd been given was correct, our fight against big tech was even more daunting than I'd ever imagined.

* * *

When the American Priorities Conference was over, Maryam and I decided to stay in Florida until the election was over, about three weeks away.

While at the conference, Maryam introduced me to one of her friends, Ian Trottier, a radio talk show host and writer, who'd worked with her on stories in 2015 regarding questions surrounding the Zika virus outbreak, which was the big scare story of that year.

Ian stands about six foot one with broad shoulders, blue eyes, and dirty blonde hair which was snow white as a kid, and looks like a bodybuilder. He loves the gym, sometimes spending eight hours working out, and has a fifty-two-inch chest. If you're going to get into a bar fight, Ian is the guy you want on your side.

The Trottier family originally came from just outside of Lyon, France, landing in Quebec in 1640. His ancestors arrived in San Francisco in 1890 and he was born in California. He got his degree from the University of Oregon, speaks fluent Spanish, and lived for many years in Miami Beach, Florida, where he started his weekly radio talk show, interviewing such well-known people as former Democratic Congresswoman Cynthia McKinney of Georgia, former *New York Times* foreign bureau chief and author of books about the CIA, Stephen Kinzer, and David Knight, one of the hosts at *InfoWars*.

Ian is an excitable guy, but when he met me his enthusiasm went off the charts. While he's the one who looks like a bruiser, he essentially designated himself as my boxing manager for my battles in the public arena.

On October 15, 2020, there was a Trump fundraiser at the Trump Doral Country Club which would feature the president, and Ian said I had to go. The only problem was the tickets cost five thousand dollars. One of the people I'd come to know since going public was Dr. Robert Epstein, a Harvard trained psychiatrist, the former editor in chief of *Psychology Today,* a visiting scholar at the University of California, San Diego, and founder of the Cambridge Center for Behavioral Studies in Concord, Massachusetts. Epstein was a political liberal, but had become the country's foremost academic voice arguing that the Big Tech platforms were manipulating the public. In a June 16, 2019 hearing before the Senate Judiciary Committee, Epstein testified that, based on his research, Google could manipulate between 2.6 million and upwards of fifteen million votes in the upcoming 2020 election.[1] Epstein had become a friend and mentor to me, and I was planning to do some work with him for a non-profit foundation. The legal paperwork hadn't yet been completed, but I called and asked if he could personally send me the money I needed.

He agreed, and I purchased the ticket.

It was billed as a fundraising dinner, but I didn't realize how early everything would start. At 2:29 p.m., I sent a quick email to the organizers, asking what time the event started, as it wasn't on the ticket, or any of the materials I scanned online.

I immediately got an email back saying the doors opened at 2:30 p.m. and would only be open for an hour, after which time nobody else would be admitted.

After the American Priorities Conference, we'd checked out of the Trump Doral and were staying at a hotel about thirty minutes away. I raced up to my room to throw my clothes on. Maryam was still in the shower, and sadly, I had to leave her behind. Ian and I had befriended another guy, Andrew, who'd been driving us around.

The three of us jumped into Andrew's car, plugged in the address of the Trump Doral Hotel, and the computer said we would arrive at 3:31 p.m. "Andrew," I said. "You've got to drive like a madman. We need to make up time. You've got to dodge in and out of cars to get ahead of everyone."

Andrew nodded and focused on his driving.

Ian suggested I contact the person I'd purchased the ticket from and tell her I was going to be there right at 3:30 p.m. She texted back and said, "Perfect," which calmed me down.

Maybe the 3:30 p.m. cutoff wasn't a hard time.

As I started to relax and we were racing to the event, Ian started hammering me on what I was going to say when I met Trump. He told me I was going to have a maximum of thirty seconds to make an impression.

I thought for a moment. "How about this? Hello, my name is Zach Vorhies, and I'm known as the Google whistleblower. I want to partner with you to topple Google and restore the First Amendment for the American people." I looked at Andrew and Ian. "That sounds too grandiose, doesn't it?"

They both thought it was perfect.

I practiced it again and again.

We got to the Trump Doral resort at 3:28 p.m. The Secret Service was at the front gate, their bomb-sniffing dogs walking past the cars, and the person at the front looking at the list and saying, "I don't see a Zach Vorhies on the list."

Luckily, Ian was in the front passenger seat and looked at the guest list. My name was not on the typed section, but somebody had written in massive letters with a pen, "ZACH VORHIES." Ian graciously, and with boyish charm and deference, pointed out my name.

And we were through the gates. Andrew pulled up at the entrance and Ian followed me out. He kept telling me I was his "asset" and he wasn't going to rest until he knew I was in the same room with Trump. We walked into the Trump Doral Country Club, me dressed in a dark suit and tie that almost looked like a tuxedo, and Ian wearing jeans, tennis shoes, a red MAGA hat, and a T-shirt.

We got directed to the reception area and I talked to the woman who was handing out badges. She looked at the computer readout and said I wasn't on the list. I told her I was a late purchase, showed her the email on my phone, and that convinced her. "And your plus one?" she asked, motioning to Ian.

Ian gave her his name and I struggled to keep myself from laughing. A minute or two later both of us had our badges. We immediately went to the bathroom, checked to see that nobody was around, and then high-fived each other and started laughing.

"Okay, we've got one job," said Ian, after we calmed down. "We have to work that room and get to Trump's inner circle. That's our mission."

We made our way into the grand ballroom where they were serving hors d'oeuvres and drinks and sushi and immediately started talking to every single person we could. I'm good at working a room, but Ian was just fantastic, going at it like an attack dog. He'd start talking me up, then bring me over and people wanted to have their pictures taken with me, gave me their business cards, and exchanged contact information.

The event was getting started and we were being told to take our seats. The front rows were reserved for VIPs and we wanted to get as close to that section as possible, but couldn't find two together. I took the closest seat to the VIP section and Ian took one further back in the center, so he'd be directly in Trump's line of sight.

I was striking out with the VIPs, getting snubbed by Lieutenant Governor of Florida Jeanette Marie Nunez, as well as Richard Grenell, the former US ambassador to Germany and from February to May of 2020 served as the acting director of National Intelligence. Maybe he was tired, but Grenell turned away when I talked to him.

As the crowd settled in their seats and the program began, Kimberly Guilfoyle, the former Fox News host and girlfriend of Don, Jr. came out and spoke. After that, Vice-President Mike Pence came out and gave a talk.

Then "Hail to the Chief" started playing and out walked President Donald J. Trump.

Trump is taller than you expect him to be, and well, you've all seen him so much, you know what he looks like. But there was something about being in that room with him, maybe it's just the aura of those in power, but it felt like the energy shifted to the top of the dial. I was just a few feet away from the most powerful man in the world.

Trump started talking, going over the polls, how he was doing in the polls, and how their own polling was showing such a different story. He talked about how he was going to ramp up his appearances to four or five times a day, how he's always fighting for us, and then he started talking about the corruption in Big Tech. He ended by talking about Big Tech, the audience started applauding, he turned to leave, and then Ian sprang to his feet.

"This is the Google whistleblower, here! Zach Vorhies, the Google whistleblower is here!" he shouting, pointing over at me.

Trump was almost off the stage, but the VIPs were moving a little more slowly, when Ian spotted Tiffany Trump (Trump's daughter with his second wife, former Miss America Marla Maples) standing with Kimberly Guilfoyle and beginning to exit.

Ian started running toward the front, shouting, "Tiffany! Tiffany!"

Tiffany turned, seemed to consider whether to talk to Ian, but Kimberly Guilfoyle motioned for her to keep going, which she did. Tiffany was quickly out of the room, the Secret Service close behind her.

It seemed we'd lost our chance.

The A players had left the room.

But a lot of the VIPs were still sitting in their seats. There was one guy, sitting next to Richard Grennel, who didn't look clean shaven, and wasn't wearing a suit, either. I noticed he was wearing a cool pair of shoes that looked like they had hollow soles.

I said to the scruffy guy, "Hey, man, those are really interesting shoes."

He looked at my shoes, five hundred dollar shoes with elaborate writing on the side that looked like a cross between a boot and a high heel tennis shoe. They were leather, green, and looked very cool. "Oh yeah, same with you, dude."

He offers a hand and says, "My name is Justin."

"I'm Zach. I'm known as the Google whistleblower."

Turns out Justin was Justin Galloway, whose family used to own the New York Yankees, and is a close childhood friend of Tiffany Trump. His eyes grew large when I told him I was the Google whistleblower and after a brief moment said, "What? I know exactly who you are. Oh, my God, you're right here!"

I said, "I'm trying to meet with President Trump because I want to partner with him to topple Google and restore the First Amendment for the American people."

He took my contact information and then said, "Stay right here. I have to do something. I'll be right back."

About five minutes later he sent me a group text.

I got a little smile on my face and walked over to Ian, who for some reason was being filmed by a woman with her cellphone. I went up to him and said, "I want you to keep taping while he reads this." I held the phone up to him.

Ian began to read, "Tiffany Trump, meet Zach Vorhies. We have a lot in common. Let's speak later if possible."

Ian's eyes got big and his mouth hung open and he arched his back to the sky like Tim Robbins in *The Shawshank Redemption*, when his character finally escaped from prison. "What the hell?"

I took the phone back and tapped out a reply to Tiffany. "Hey, Tiffany Trump, my name is Zach Vorhies. I'm known as the Google whistleblower. One year ago, I resigned and released 950 pages of internal documents revealing their intensive censorship regime to the DOJ and the public with Project Veritas. I did this to warn the public of the coming election coup in 2020. I want to partner with the President to topple Google and restore the First Amendment for the American public."

She texted me back. "That is exactly what I focused on at law school. [She had just graduated from the University of Pennsylvania.] Then I worked closely with Dr. Robert Epstein and Charlene Bollinger [Co-producer of the documentaries *The Truth About Cancer* and *The Truth About Vaccines*] told me she interviewed you."

"I love Charlene and Robert is a close friend." He was also the guy who granted me five thousand bucks so I could go to the event.

Tiffany texted back, "Of course."

Justin came back and found me. "We're going to meet with Tiffany Trump," he said. "Just not here. We're going to go to another room."

I grabbed Ian and we followed Justin. The A-listers were hanging out in another room at the club and a few turned to look at us as we entered. First was Kimberly Guilfoyle, coming up to talk to me and Ian. Here we were talking to this mega-celebrity, just trying to act all calm and collected. Others came up and started talking to us, many of them asking for pictures with me.

The doors to an adjoining conference room opened up and out walked an entourage of people with Tiffany Trump in the center, heading toward me like a speeding bullet.

She thanked me for what I'd done and we started talking about our mutual friends, Dr. Robert Epstein and Charlene Bollinger. Maybe because she knew I was friends with Charlene, she said President Trump would never mandate a vaccine. I thanked her for that and then in a rush told her about Google planning the election coup and how I'd sacrificed everything to get the information out to the American people. I gave her some ideas on how we might fight Google and she said she was interested.

We spoke for a few moments, she said it was great, but she had to run off to the Presidential Town Hall for her father. She asked if I'd be back later and I said I would.

And with that, Tiffany Trump was gone.

* * *

I needed to decompress. So much had happened during the day, from not knowing if we'd make it to the event, to getting to meet the president's daughter, who was a fan of mine!

She made me feel like I was the celebrity!

Justin suggested we hang out in his room, and some other people were there, including Professor Gerald Protheroe, an author and an assistant associate professor at New York University, and his wife, Rita, who worked in Washington, DC and had great stories to tell. We smoked cigars on the balcony, drank, and just had a great time.

After a few hours, we went down to the famed bar at the Trump Doral, which overlooks the golf course. We had been there about an hour when Secret Service started coming in, checking out the room, and somebody announced, "Ladies and gentlemen, the president of the United States!"

Trump walked in, Tiffany next to him, and he scanned the room, acknowledging the applause of the group in the bar.

Then Trump pointed directly at me. "You, sir, you're an American hero," he said. "What you did, you did a really great job for this country. Thank you so much. The Google whistleblower, right here. Great job."

Ian started going crazy, yelling and clapping with his hands above his head.

Trump turns his attention to Ian, looking amused. "Who's this guy? He looks like a prizefighter."

Of course, Ian just loved that. From now to the end of time, Ian will refer to himself as the "prizefighter."

Trump then turned his attention back to me. "What did you think about Joe Biden's thing tonight?" Joe Biden had a dueling town hall on CBS that night.

"Who?" I asked, trying to make a joke.

"Joe Biden," said the president.

I said, "Oh, who cares about him? He's a nobody. You did fantastic, Mr. President."

EPILOGUE

Where We Go From Here

The Presidential election of November 3, 2020 was a catastrophic failure for the United States.

Not because the official election results showed Joe Biden beat Donald Trump, but because a significant number of Americans do not believe the result. *Rasmussen Reports* published an article on February 12, 2021, about three weeks after Joe Biden had been sworn in, which showed the country was still deeply divided:

> The percentage of voters who believe Biden won the 2020 election fairly has not increased much since early January. At that time, 55% of voters said Biden had been elected fairly, 39% said he had not. In that January survey, 69% of GOP voters said Biden had not won the election fairly . . .
>
> Republicans still overwhelmingly believe mail-in voting resulted in fraud. Seventy (70%) of GOP voters say mail-in voting led to unprecedented fraud in the 2020 election, as do 11% of Democrats and 46% of voters not affiliated with either major party.[1]

It strikes me as a highly volatile situation in our country when a significant number of Republicans, Democrats, and Independents do not view the election results with confidence. In this book, I do not make the claim our recent election had integrity or lacked it. I am just as much in the dark as anyone about these questions.

However, given the apparent fact that 70 percent of Republicans, 46 percent of Independents, and 11 percent of Democrats believe there was unprecedented fraud in the election, I am surprised why our leaders on both sides of the aisle are not rushing to the public with detailed information showing it was a fair election. It seems to me there are only two possibilities. The first is that the claims are so ridiculous that our leaders feel free to ignore our concerns. The second is that the establishments of both parties have been engaging in similar shenanigans for years, and they don't want that information becoming public. I am unable to come up with any other possible explanations.

In this story, I've detailed my experiences at Google, and how I believe they developed an information control system with the intent of defeating President Trump. I don't believe I've gone beyond the evidence I've described. And when I've speculated, I've clearly stated it as speculation.

However, it seems to me that as we are getting more information, we are getting less transparency as to how such decisions are being made, and less alternative opinions about what it all means. My publication of the Google documents was meant to stimulate a conversation about how our understanding of the world is being shaped by a relatively small number of people in Big Tech.

Some will see in my story a direct link between Google's attempt to create a system of information control (rather than just cataloguing the world's information), and the defeat of President Trump for a second term. For others, they will claim my story doesn't mean that at all. Others may claim I was successful in showing a left-wing bias at Google, but it doesn't matter because companies are free to engage in politics, and if the right wants to build a conservative Google, they're free to do so. I am not responsible for what people take away from my story. I will remain agnostic on what it all means.

I did what I could in the run-up to the 2020 election and do not regret any of it, despite whether or not those closest to me understand my motivations.

I believe we are living in the middle of history, not the end. Each of us must do our part to bring about a better tomorrow.

I have done what I can to tell the truth of what I saw at Google and YouTube. Others are stepping up to do their part as well. The slogan of Project Veritas, "Be brave. Do something," is becoming the operating code of many concerned about this country's future. Each generation's fight for freedom will look a little different.

This war is being fought on a digital battlefield. Instead of tanks, rockets, and bombs dropped from the air, we have lawyers, the media, and paid off politicians bombarding us with their lies.

But people are innovators, especially when the scope of a problem becomes clear to them. I hope I've shed some light on these issues and that my fellow engineers are working right now on solutions. I know I am.

When people ask me, "Zach, what do you think is going to come of all this?" I always say the same thing.

"Something wonderful."

AFTERWORD

How to Defeat Censorship

How do you defeat censorship?

This is a question I get asked a lot, and I have a very easy answer.

"Aggregation cancels censorship."

What exactly made YouTube great? Well YouTube is two things: (1) It's a video hosting platform and (2) It's a video recommendation engine. This works great when everyone can be in the same platform.

But then the YouTube purge happened on October 15, 2020, and many of the conservative content creators got banned. So where exactly did they go? Of 154 content creators that I tracked, 75 percent of them were still on YouTube, while Rumble had 13 percent and Bitchute had 11 percent (as of March 2021).

The problem now is that conservative content is fragmented. Users now have to play the "social media shuffle" to get to each of the conservative video creators every day. This turns out to be hours of work for the highest news consumers. Most casual news consumers won't go to great lengths to find the content they previously liked—which is why censorship works so well!

So why did YouTube censor content? It's been my experience that Conservative and Anti-Globalist dissent has high engagement, whether measured in clicks, likes, comments, or whatever, conservative content trends naturally. YouTube tried to tweak their recommendation engine but failed so they just started banning everyone. This game of whack a mole caused the YouTube recommendation algorithm to then over-recommend anyone left standing. This is why YouTube had to do the purge. Their AI recommendation engine was causing too many people to see conservative content and creating a mass shift in consciousness.

After the YouTube purge, what did we miss the most? The answer is that we miss the Recommendation Engine. The alt-tech platforms can't fix this because they are limited to the content on their own systems! But why does aggregation have to be tied to hosting video content? It doesn't!

For example, if Scott Adams is censored on YouTube, would he stop producing videos? OF COURSE NOT! He's an intellectual with a crippling social media addiction—of course he's going to continue making videos! So a hypothetical aggregator can just switch from pulling videos from Scott Adam's YouTube account to say Rumble and continue serving his videos.

This is more than just talk though, this is something that I'm working on right now and soon I will launch.

So, in the end, I ended up right where I began. In San Francisco, working on a video platform so people like you can connect to people you like and see the videos you love. But this time, a wiser and less naive person sits at the helm. One without the ill-conceived notion that a large mega corporation has America's interest at heart. The question is, if we build our alternative platforms, how far will big tech go to try and stop us?

Notes

Epigraph

1. David McCabe, Cecilia Kang, and Daisuke Wakabayashi, "U.S. Accuses Google of Illegally Protecting Monopoly," New York Times, October, 20, 2020, www.nytimes.com/2020/10/20/technology/google-antitrust.html, link to full suit: www.nytimes.com/interactive/2020/10/20/us/doj-google-suit.html.

Chapter One: Google Turns toward the Dark Side

1. George Washington, "In 1967, the CIA Created the Label "Conspiracy Theorists" . . . to Attack Anyone Who Challenged the 'Official' Narrative," *Zero Hedge*, February 23, 2015, www.zerohedge.com/news/2015-02-23/1967-he-cia-created-phrase-conspiracy-theorists-and-ways-attack-anyone-who-challenge.
2. Julie Bort, "An Inside Look at Google's Luxurious 'Googleplex' Campus in California," *Financial Post*, October 7, 2013, www.financialpost.com/business-insider/an-inside-look-at-googles-luxurious-googleplex-campus-in-california.
3. Samantha Todd, "Google Head of HR Eileen Naughton to Step Down Amid Unrest at Tech Titan," *Forbes*, February 11, 2020, www.forbes.com/sites/samanthatodd/2020/02/11/google-head-of-hr-eileen-naughton-to-step-down-amid-unrest-at-tech-titan/#312176c04b91.

Chapter Five: Building the "Ministry of Truth"

1. Jeff Dunn, "Facebook's Fake News Problem in One Chart," *Business Insider*, November 18, 2016, www.businessinsider.com/facebook-fake-news-donald-trump-buzzfeed-chart-2016-11.
2. Ibid.
3. Tim Haines, "FLASHBACK: Rand Paul Asks Hillary Clinton About Arms Smuggling from Benghazi to ISIS," Real Clear Politics, May 19, 2015, www.realclearpolitics.com/video/2015/05/19/flashback_rand_paul_asks_hillary_clinton_about_arms_smuggling_out_of_libya.html.

4. Rich Buhler, "Clinton Sent Ambassador Stevens to Benghazi to Retrieve Stinger Missiles—Unproven!" Truth or Fiction, October 18, 2016, www.truthorfiction.com/clinton-sent-ambassador-stevens-benghazi-retrieve-stinger-missiles/.
5. Roger L. Simon, "Exclusive: Ex-Diplomats Report New Benghazi Whistleblowers with info Devastating to Clinton and Obama," PJ Media, May 21, 2013, www.pjmedia.com/rogerlsimon/2013/05/21/pjm-exclusive-ex-diplomats-report-new-benghazi-whistleblowers-with-info-devastating-to-clinton-and-obama-n218228.
6. Ibid.
7. Plato, "The Republic," Translation by G.M.A. Grube, Revised by C.D.C. Reeve, Hackett Publishing Company, (Indianapolis, Indiana), 1992, p. 68.
8. Project Veritas, "Google 'Machine Learning Fairness' Whistleblower Goes Public, says 'Burden Lifted Off My Soul,'" August 14, 2019, www.projectveritas.com/news/google-machine-learning-fairness-whistleblower-goes-public-says-burden-lifted-off-of-my-soul/, Document found at: www.pv-uploads1.s3.amazonaws.com/uploads/2019/06/SS1DocDump.pdf.
9. Ibid.
10. Ibid.
11. Ibid.
12. Ibid.
13. Ibid.
14. Ibid.
15. Ibid.
16. Ibid.
17. Ibid.

Chapter Six: The Covfefe Deception

1. Ibid.
2. Liam Stack, "No, 'Covfefe' was not Trump Speaking Arabic," *The New York Times*, May 1, 2017, www.nytimes.com/2017/06/01/us/politics/covfefe-trump-arabic.html.
3. Ibid.
4. Ibid.
5. "'Covfefe' Translate Easter Egg," Google Internal Document, Created June 1, 2017, www.zachvorhies.com/covfefe/1.pdf.
6. Covfefe Discussion Thread, June 1,2017 to June 5, 2017, www.zachvorhies.com/covfefe/2.pdf.
7. "Jacques Derrida," Stanford Encyclopedia of Philosophy, (Published November 22, 2006, revised July 30, 2019), www.plato.stanford.edu/entries/derrida/.

8. Elizabeth Powers, "'Woke' Culture: 1970s Deconstruction Goes Mainstream," *National Review*, June 29, 2020, www.nationalreview.com/2020/06/woke-culture-1970s-deconstruction-goes-mainstream/.
9. Covfefe Discussion Thread, June 1,2017 to June 5, 2017, www.zachvorhies.com/covfefe/2.pdf.
10. Ibid.
11. Ibid.
12. Ibid.
13. Ibid.
14. Joel Goldstein, "Trump Opponents Have Rediscovered the 25 Amendment. Here is What You Should Know About it," The Washington Post, June 7, 2017, www.washingtonpost.com/news/monkey-cage/wp/2017/06/07/5-things-you-should-know-about-the-25th-amendment/.
15. Ibid.
16. Ibid
17. Anonymous, "I Am Part of the Resistance Inside the Trump Administration," *The New York Times*, September 5, 2018, www.nytimes.com/2018/09/05/opinion/trump-white-house-anonymous-resistance.html.
18. Aaron Blake, "Did Top Trump Administration Officials Suggest Removing Him from Office? It's Still as Clear as Mud," *The Washington Post*, June 3, 2020, www.washingtonpost.com/politics/2020/06/03/did-top-trump-administration-officials-seek-remove-him-office-its-still-clear-mud/.
19. Robert Lewis, "Susan Wojcicki, American Tech Industry Executive," Encyclopedia Britannica, (Accessed October 4, 2020), www.britannica.com/biography/Susan-Wojcicki.
20. Ibid.
21. Laura Lorenzetti, "Google's Sergey Brin and 23and Me's Anne Wojcicki Legally Divorced, June 24, 2015, www.fortune.com/2015/06/24/google-sergey-brin-anne-wojcicki-divorce/.
22. Ibid.
23. Robert Cole, "Are Gavin Newsom and Nancy Pelosi Related?" Celeb Answers, September 10, 2020, www.celebanswers.com/are-gavin-newsom-nancy-pelosi-related/.
24. Ibid.
25. "YouTube CEO Susan Wojcicki Explaining News Editorial Decisions 'Authoritative" and 'Fake' News," YouTube, August 17, 2019, www.youtube.com/watch?v=ZuVM-kG5tQw.
26. Ibid.
27. Patsy Widakuswara, "Trump Hosts Abraham Accords and Signing Between Israel, UAE, and Bahrain," Voice of America, September 15, 2020, www.voanews.com/middle-east/trump-hosts-abraham-accords-signing-between-israel-uae-and-bahrain.

Chapter Seven: The Las Vegas Massacre

1. History.com editors, "Gunman Opens Fire on Las Vegas Concert Crowd, Wounding Hundreds and Killing 58," This Day in History, History Channel, (October 1, 2018, updated September 29, 2020), www.history.com/this-day-in-history/2017-las-vegas-shooting.
2. Ibid.
3. Ibid.
4. "Las Vegas Shooting: A Timeline of the Police Response to the Attack," ABC News, October 4, 2017, updated October 9, 2017, www.abc.net.au/news/2017-10-04/las-vegas-shooting-a-timeline-of-events/9015258.
5. Sabrina Tavernise, Serge F. Kovaleski, and Julie Turkewitz, "Who Was Stephen Paddock? The Mystery of a Nondescript 'Numbers Guy,'" *New York Times*, October 7, 2017, www.nytimes.com/2017/10/07/us/stephen-paddock-vegas.html.
6. Ibid.
7. Ibid.
8. Edna Boykin, "Money Laundering and Casinos: Things You Should Know," Times of Casino, June 9, 2020, www.timesofcasino.com/money-laundering-and-casinos-things-you-should-know/.
9. Sabrina Tavernise, Serge F. Kovaleski, and Julie Turkewitz, "Who Was Stephen Paddock? The Mystery of a Nondescript 'Numbers Guy,'" *New York Times*, October 7, 2017, www.nytimes.com/2017/10/07/us/stephen-paddock-vegas.html.
10. Ibid.
11. Ibid.
12. Todd Prince, "Mandalay Bay Renumbering Floors Associated with Las Vegas Shooting," *Las Vegas Review Journal*, February 6, 2018, www.reviewjournal.com/crime/shootings/mandalay-bay-renumbering-floor-associated-with-las-vegas-shooting-1304543/.
13. Anupreeta Das and Craig Karmin, "Two VIP Billionaires Teamed Up to Run Luxury Hotels. It's Been a Slog," *Wall Street Journal*, July 16, 2017, www.wsj.com/articles/two-vip-billionaires-teamed-up-to-run-luxury-hotels-its-been-a-slog-1500226911.
14. Ibid.
15. Ibid.
16. Ibid.
17. Martin Chulov, "I Will Return Saudi Arabia to Moderate Islam, says Crown Prince," *The Guardian*, October 24, 2017. www.theguardian.com/world/2017/oct/24/i-will-return-saudi-arabia-moderate-islam-crown-prince.
18. Ibid.
19. MGUR, "MAGA-Truth: Evidence Volume I: Lock Her Up," November 12, 2017, www.imgur.com/r/The_Donald/DTeK7.

20. Lance Goodall, "The Saudi Royal Connection and the Las Vegas Concert Massacre," Coercion Code, November 10, 2017, www.coercioncode.com/2017/11/10/the-saudi-royal-connection-and-the-las-vegas-concert-massacre/.
21. Ibid.
22. "Yemen's Iran-Backed Houthi Rebels Fire Ballistic Missile at Saudi Capital of Riyadh," *Haaretz*, November 5, 2017, www.haaretz.com/middle-east-news/yemen-s-houthis-fire-ballistic-missile-at-saudi-capital-of-riyadh-1.5629085.
23. David D. Kirkpatrick, "Saudi Arabia Arrests 11 Princes, Including Billionaire Alwaleed bin Talal," *New York Times*, November 4, 2017, www.nytimes.com/2017/11/04/world/middleeast/saudi-arabia-waleed-bin-talal.html.
24. Douglas Jehl, "Buffet of Arabia," *New York Times*, March 28, 1999, www.nytimes.com/1999/03/28/business/buffett-of-arabia-well-maybe.html.
25. Ibid.
26. David D. Kirkpatrick, "Saudi Arabia Arrests 11 Princes, Including Billionaire Alwaleed bin Talal," *New York Times*, November 4, 2017, www.nytimes.com/2017/11/04/world/middleeast/saudi-arabia-waleed-bin-talal.html.
27. Ibid.
28. "Saudi Helicopter Crash 'Kills High-Ranking Prince,'" Al Jazeera, November 6, 2017, www.aljazeera.com/news/2017/11/6/saudi-helicopter-crash-kills-high-ranking-prince.
29. "Sources: Saudi Prince's Helicopter was Targeted by Bin Salman," Middle East Monitor, November 8, 2017, www.middleeastmonitor.com/20171108-sources-saudi-princes-helicopter-was-targeted-by-bin-salman/.
30. "Blacklist - Search Burganizer Las Vegas Massacre," Internal Google Document, (Created October 2, 2017, 7:31 am, Pacific Time), www.zach-vorhies.com/blacklists/page_level_domain_blacklist.pdf.
31. Ibid.
32. Ibid.
33. Ibid.
34. Ibid.
35. Ibid.
36. Ibid.
37. Ibid.
38. Ibid.
39. Ibid.

Chapter Nine: Approaching *Breitbart*

1. "Acorn—Baltimore—Part 1," Project Veritas, (Accessed October 18, 2020), www.projectveritas.com/investigation/acorn/.
2. Ibid.
3. Ibid.
4. Ibid.

5. Ibid.
6. "Deconstructing Race: Analyzing Inequities in a Racial Society," Talks at Google, (December 7, 2016), www.talksat.withgoogle.com/talk/deconstructing-race-analyzing-inequities-in-a-racial-society.
7. The Edge 45 Team, "The Google E-A-T Score: What is it and Why Does It Matter," Better Marketing, June 5, 2019, www.medium.com/better-marketing/the-google-e-a-t-score-what-is-it-199f889f756a.
8. Ian Booth, "E-A-T and SEO: How to Create Content Google Wants," *Moz*, June 4, 2019, www.moz.com/blog/google-e-a-t.
9. Charlie Nash, "Twitter Engineer Admits to Banning Accounts that Express Interest in God, Guns, and America," *Breitbart*, January 11, 2018, www.breitbart.com/tech/2018/01/11/twitter-engineer-admits-to-banning-accounts-that-express-interest-in-god-guns-and-america/.

Chapter Ten: The Youtube Shooter

1. Ibid.
2. "Latest: Police: YouTube Shooter was Angry with the Company," *Associated Press*, April 4, 2018, www.apnews.com/article/3c7250eb9d824a5b921939eb4c82fb1e.
3. "Nasim Name Meaning," *Urdu Point*, (Accessed October 24, 2020), www.urdupoint.com/islamic-names/nasim-name-meaning-in-english-90850.html.
4. "Latest: Police: YouTube Shooter was Angry with the Company," *Associated Press*, April 4, 2018, www.apnews.com/article/3c7250eb9d824a5b921939eb4c82fb1e.
5. Sudin Thanawala and Janie Har, "Police Plan Lengthy Investigation of YouTube Shooter's Past," *Seattle Times*, April 4, 2018, updated April 5, 2018, "Latest: Police: YouTube Shooter was Angry with the Company," *Associated Press*, April 4, 2018, www.seattletimes.com/nation-world/youtube-shooters-bizarre-videos-key-to-suspected-motive/.
6. "Latest: Police: YouTube Shooter was Angry with the Company," *Associated Press*, April 4, 2018, www.apnews.com/article/3c7250eb9d824a5b921939eb4c82fb1e.
7. Rachel Blevins, "5 Compelling Reasons the YouTube Shooting Disappeared from the Headlines," The Free Thought Project, April 7, 2018, www.thefreethoughtproject.com/5-compelling-reasons-why-the-youtube-shooting-has-disappeared-from-headlines/.
8. Ibid.
9. Ibid.
10. Ibid.
11. Ibid.
12. Ibid.

Chapter Eleven: Project Veritas Returns

1. Christian Hartsock, "Insider Blows Whistle & Exec Reveals Google Plan to Prevent "Trump Situation" in 2020 on Hidden Camera," *Project Veritas*, June

24, 2019, www.projectveritas.com/video/insider-blows-whistle-exec-reveals-google-plan-to-prevent-trump-situation-in-2020-on-hidden-cam/.

2. Jen Gennai, "This is Not How I Expected Monday to Go!" *Medium*, June 24, 2019, www.medium.com/@gennai.jen/this-is-not-how-i-expected-monday-to-go-e92771c7aa82.
3. Text Message to coauthor Kent Heckenlively, October 30, 2020.
4. Christian Hartsock, "Insider Blows Whistle & Exec Reveals Google Plan to Prevent "Trump Situation" in 2020 on Hidden Camera," *Project Veritas*, June 24, 2019, www.projectveritas.com/video/insider-blows-whistle-exec-reveals-google-plan-to-prevent-trump-situation-in-2020-on-hidden-cam/.
5. "Google Mission and Vision Statement Analysis," Google, (Accessed October 31, 2020), www.mission-statement.com/google/.
6. Christian Hartsock, "Insider Blows Whistle & Exec Reveals Google Plan to Prevent "Trump Situation" in 2020 on Hidden Camera," *Project Veritas*, June 24, 2019, www.projectveritas.com/video/insider-blows-whistle-exec-reveals-google-plan-to-prevent-trump-situation-in-2020-on-hidden-cam/.
7. Ibid.
8. Ibid.
9. Ibid.
10. Ibid.
11. Ibid.
12. Ibid.
13. Ibid.
14. Ibid.
15. Ibid.
16. Ibid.
17. Ibid.
18. Ibid.
19. Ibid.
20. Ibid.
21. Ibid.
22. Ibid.
23. "Sun Tzu Quotes—The Art of War," *Goodreads*, (Accessed November 1, 2020), www.goodreads.com/quotes/56058-let-your-plans-be-dark-and-impenetrable-as-night-and.
24. Telephone Interview with Cassandra Spencer by Kent Heckenlively, November 1, 2020.
25. Ibid.
26. Ibid.
27. "Google 'Machine Learning Fairness' Whistleblower Goes Public, says: 'Burden Lifted Off of My Soul," *Project Veritas*, August 14, 2019, www.projectveritas.com/news/google-machine-learning-fairness-whistleblower-goes

-public-says-burden-lifted-off-of-my-soul/ Original letter—www.pvuploads1.s3.amazonaws.com/uploads/2019/08/gz-google-letter-1.png.

Chapter Twelve: Reclaiming My Soul

1. "Google "Machine Learning Fairness" Whistleblower Goes Public, says: "burden lifted off of my soul," *Project Veritas*, August 14, 2019, www.projectveritas.com/news/google-machine-learning-fairness-whistleblower-goes-public-says-burden-lifted-off-of-my-soul/.
2. Ibid.
3. Ibid.
4. Mohamed A. El-Erian, "Revolution 2.0—How One Google Executive and Facebook Sparked an Uprising in Egypt," Huffington Post, May 24, 2013, updated July 24, 2013, www.huffpost.com/entry/revolution-20-how-one-goo_b_3333340.
5. Ashraf El-Sherif, "The Egyptian Muslim Brotherhood's Failures," Carnegie Endowment for international Peace," July 1, 2014, www.carnegieendowment.org/2014/07/01/egyptian-muslim-brotherhood-s-failures-pub-56046.
6. Ibid.

Chapter Thirteen: What Bravery Brings to You

1. Robert Epstein, "Why Google Poses a Serious Threat to Democracy, and How to End that Threat," United States Senate Judiciary Subcommittee on the Constitution, June 16, 2019, www.judiciary.senate.gov/imo/media/doc/Epstein%20Testimony.pdf.

Epilogue: Where We Go From Here

1. "Most GOP Voters Still Don't Think Biden was Elected Fairly," Rasmussen Reports, February 12, 2021, www.rasmussenreports.com/public_content/politics/elections/election_2020/most_gop_voters_still_don_t_think_biden_was_elected_fairly.